Radio-
Electronics'®

STATE OF SOLID STATE

By the Editors of Radio-Electronics

TAB
TAB BOOKS Inc.
Blue Ridge Summit, PA 17214

FIRST EDITION

FIRST PRINTING

Printed in the United States of America

Library of Congress Cataloging in Publication Data

Radio-electronics' state of solid state.

Includes index.
1. Solid state electronics. I. Radio-electronics.
II. Title: State of solid state.
TK7871.85.R34 1987 621.3815'3 87-10190
ISBN 0-8306-7733-X
ISBN 0-8306-2733-2 (pbk.)

Questions regarding the content of this book should be addressed to:

Reader Inquiry Branch
Editorial Department
TAB BOOKS Inc.
P.O. Box 40
Blue Ridge Summit, PA 17214

Cover photograph courtesy of Siliconix.

Contents

Introduction

This book is a compilation of several years of the article "State of Solid State," from *Radio-Electronics* magazine. It contains articles from February of 1982 to December of 1985.

In these articles you will find a unique mix of semiconductor theory, information on various solid state devices, and projects using those devices. We hope that you will find them very useful.

Chapter 1

SCR Arrays and Bar-Graph Displays

Let's begin with a look at two new SCR arrays from Sprague. They are the UTN-2886B and the UTN-2888A monolithic multiple-SCR arrays designed for coupling microprocessors, demultiplexers, and similar devices to high-current loads such as relays, solenoids, and lamps. The devices feature low input-current; TTL, LS-TTL, and CMOS compatibility; 35-volts minimum forward-blocking-voltage, and a 2-amp in-rush-current capability. They can be used with either full-wave or half-wave power sources.

The UTN-2886B consists of six individual SCRs and two pairs of paralleled SCRs (pins 1-16 and 8-9). Each separate SCR can handle 250 mA—the paralleled units at pins 9 and 16 handle 500 mA—at +50 °C ambient temperature.Power dissipation can be increased by attaching an external heat sink. The UTN-2888A array contains eight isolated SCRs (see Fig. 1-1), each capable of continuous and simultaneous operation at +50 °C.

Characteristics of any individual SCR (maximum limits are given unless otherwise specified) are: Forward blocking-current, 50 μA; gate-to-anode leakage, 250 μA; forward ON-voltage, 1.2 volts at +25 °C and 1.15 volts at +55 °C; gate trigger-voltage, 2.5 volts; gate trigger-current, 300 μA; gate OFF-current, 10 μA; holding current, 10 mA at 0 °C and 5.0 mA at +55 °C, and anode OFF-voltage, 400 mV minimum.

The devices operate from unfiltered DC from half-wave or full-wave-rectified sources. They cannot be used with raw AC. In

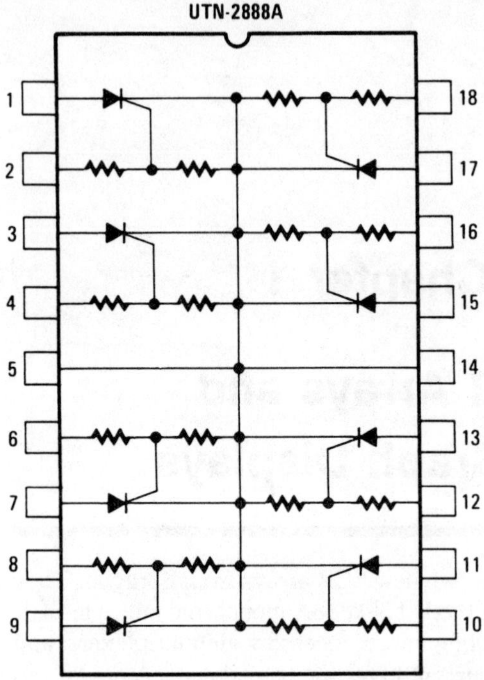

Fig. 1-1.

operation, the SCR is triggered ON by applying a positive voltage to its gate. It continues to conduct, even though the gate voltage is removed or made slightly negative, until the anode-cathode voltage is reduced to below 400 mV.

The diagram in Fig. 1-2 shows a typical application, using the SCRs as lamp drivers. When using multiple SCRs and a common V_{AA} supply, gate-to-anode leakage currents (I_{GA}) can hold the supply voltage above the anode OFF voltage, $V_{AK(OFF)}$, and prevent proper turn-off. Resistor R is connected across the power supply to ensure proper SCR operation.

The maximum value of the resistor can be calculated from the following formula:

$$\frac{400 \text{ mV}}{(n-1)\ 250\ \mu\text{A}}$$

where n is the number of SCRs being used in the system. Be sure to add 2 to the SCR count if you use the paralleled devices connected to pins 9 and 16 of the UTN-2886B.

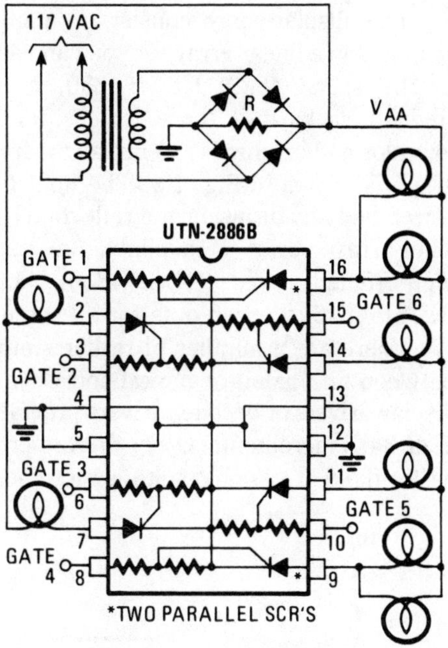

Fig. 1-2.

BAR GRAPH CIRCUIT

Bar-graph displays are popular as replacements for analog meters in radios, audio equipment, etc. The bar-graph circuits are simple and easily adapted for use with existing equipment. The analog signal needs little processing before being fed to the bar-graph display.

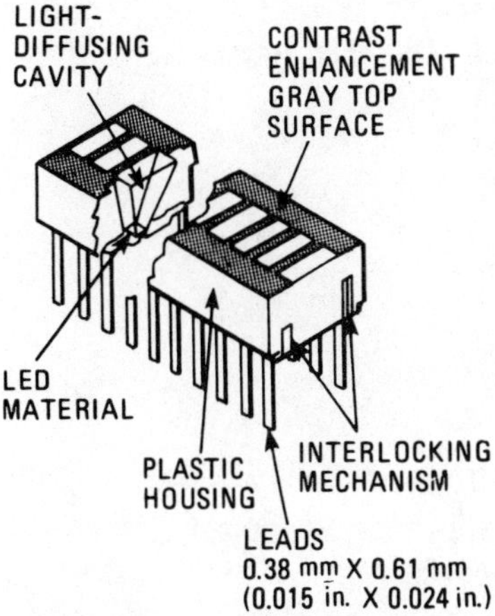

Fig. 1-3.

The display may consist of a number of discrete LEDs, arranged in a linear array, or you can use a 10-element device such as the HDSP-4820, HDSP-4830, or HDSP-4840 series from Hewlett-Packard. In those devices, the LEDs have been matched for color and brightness. A cut-away drawing of a 10-element bar-graph is shown in Fig. 1-3. The light from each LED element is "stretched" by diffusion and reflection to form individual elements. The array series is available in standard red (HDSP-4820), high-efficiency red (HDSP-4830), and high-efficiency yellow (HDSP-4840); the pin-outs for all three are the same.

There are a number of rather simple display-drivers for use between an analog or digital source and the LED arrays. Some display drivers or decoder/drivers provide a logarithmic display with 2- or 3-dB increments. Other drivers provide a linear display with either fixed or variable voltage increments. The drivers are made

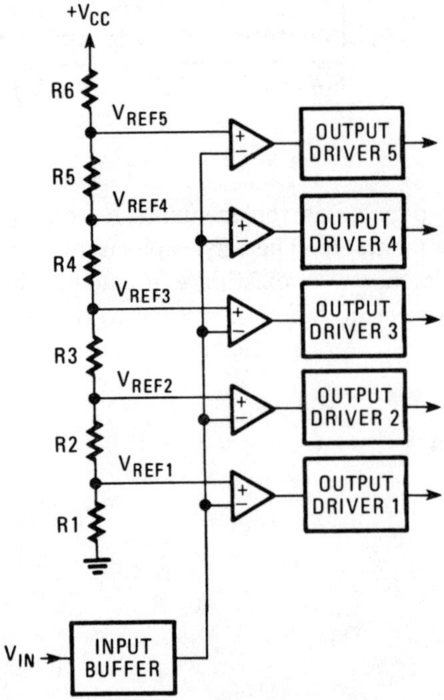

Fig. 1-4.

$V_{IN} > V_{REF1}$ OUTPUT DRIVER 1 ENABLED
$V_{IN} > V_{REF2}$ OUTPUT DRIVERS 1,2 ENABLED
$V_{IN} > V_{REF3}$ OUTPUT DRIVERS 1,2,3 ENABLED
$V_{IN} > V_{REF4}$ OUTPUT DRIVERS 1,2,3,4 ENABLED
$V_{IN} > V_{REF5}$ OUTPUT DRIVERS 1,2,3,4,5 ENABLED

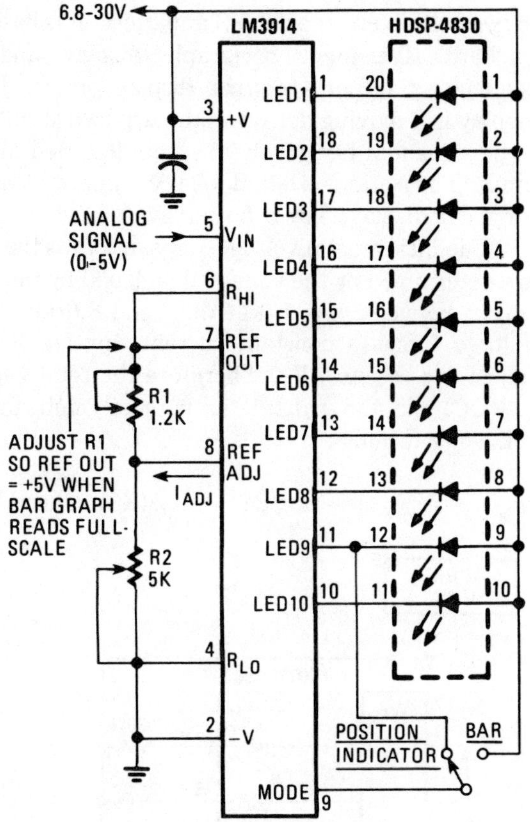

Fig. 1-5.

by, among others, National, Texas Instruments, Exar, Siemens, and AEG-Telefunken.

A typical bar-graph decoder consists of a set of voltage comparators and a precision voltage-driver that supplies reference voltages to the individual comparators. Figure 1-4 shows a representative analog-input 5-stage bar-graph-display decoder. The decoder drives the display in a logarithmic or linear manner, depending on the scaling of the reference-voltage network.

Some decoders provide displays in either bar or dot-position modes. In the bar mode, the decoder lights all the LEDs whose comparator voltage-thresholds are below the level of the applied input signal. In the dot-position mode, the display shows a moving dot whose positive is determined by the highest-threshold-voltage comparator triggered by the applied signal voltage.

Figure 1-5 shows a circuit that displays signal levels from zero to five volts in either a dot or bar pattern. It is based on the Hewlett-Packard 10-element bar-graph display and National Semiconductor's LM3914 dot/bar display driver. In this circuit, the display is a moving dot when the driver's MODE pin (pin 9) is tied to pin 11 and a bar graph when pin 9 is tied directly to $+V_{CC}$ at pin 3. (The National Data Book recommends leaving pin 9 floating for a dot display. Try it both ways.)

The incremental voltage-steps and thus the full-scale voltage, are determined by the voltage developed by the internal reference source between pins 7 (REF OUT) and 8 (REF ADJ). That constant voltage forces a constant current through R1 and through an adjustable resistor, R2. Therefore the total voltage between the REF OUT pin and ground can be easily determined by using the following formula:

$V_{OUT} = V_{REF}\,(1 + R2/R1) + I_{ADJ}(R2)$: where $V_{REF} = 1.25$ volts

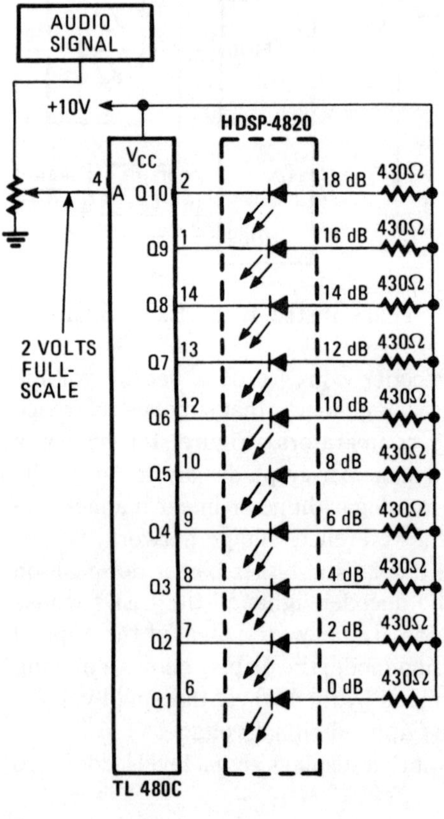

Fig. 1-6.

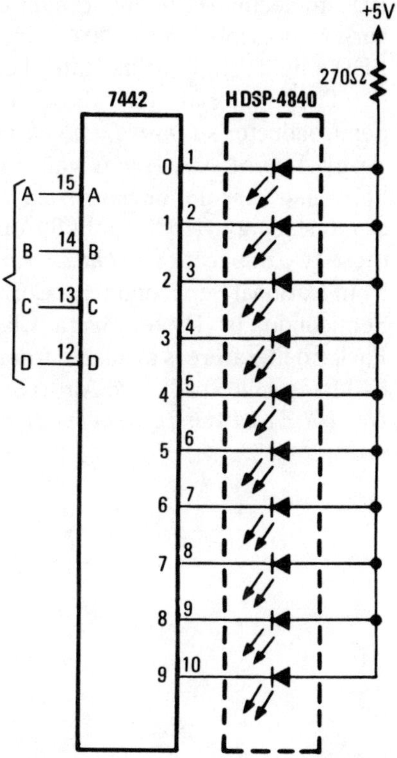

Fig. 1-7.

and I_{ADJ} is 75 μA typical, 150 μA maximum.

The current drawn from the voltage-reference pin determines the LED current. Approximately 10 times this current passes through each lighted LED. The LED current, and thus the brightness, can be adjusted by resistor R2. With a 7-volt supply, the LM3914 dissipates approximately 110 mW in the DOT mode, and around 720 mW in the BAR mode.

A low-cost VU meter is shown in Fig. 1-6. The bar-graph display is driven logarithmically by a Texas Instruments TL480C 10-step logarithmic analog level-detector. The device uses ten comparators to detect the level of the analog signal applied to the analog input (pin 4). Output Q1 is switched to a logic-low, turning on the 0-dB indicator LED when the input signal reaches approximately 218 mV. The comparators switch in 2-dB increments until they are all low at an input value of approximately 1.732 volts.

In some systems, a BCD (*B*inary *C*oded *D*ecimal) signal is the most convenient way to drive a bar graph. Figure 1-7 shows a 7442

BDC-to-decimal (1 to 10) decoder connected to drive a 10-element bar-graph display in the DOT mode. In the case of that circuit, the LED's light singly to indicate the level of the BCD signal.

The material for these circuits came from National Semiconductor's *Linear Databook* and Hewlett-Packard's *Bar Graph Array Applications*. If you want to see other applications such as "chaining" for dot or bar displays of 20 or more LED elements, or interfacing to a 6800 or 8080A microprocessor, we recommend those two publications. *The Linear Databook* is available for $9.00 from National Semiconductor Corp., Literature Distribution, 2900 Semiconductor Drive, Santa Clara, CA 95051. The Hewlett-Packard literature is available from Hewlett-Packard Components, 640 Page Mill Road, Palo Alto, CA 94304; ask for *Application Note No. 1007* and the *10-Element Bar Graph* data sheet.

Chapter 2

Musical ICs

Hardly a day passes that I don't read about some interesting new or improved semiconductor device, or of an innovative application for a more familiar one. But all too often, some of the most exciting devices are available only to equipment manufacturers, or are priced beyond the range of the average hobbyist.

That is not the case with a series of ICs available from Epson America. It has released a series of 24 melody-generator CMOS LSI devices. They are programmed to play selected melodies and to produce chime and alarm tones. They can be used in musical toys, to replace doorbells and chimes, as replacements for the mechanical devices in music boxes, and to produce musical tones that are a pleasant change from the strident clanging of a gong or the rasping sounds of a buzzer used in many applications. The devices are nearly complete in themselves, requiring only speaker, 1.5-volt battery, and a few external parts for operation.

The 7910 series of ICs provides a selection of two preprogrammed tunes plus a chime and an alarm tone. The melody arrangements are diphonic—two notes sound at once. The 7930 series of ICs play only a single tune. The 7920s are simple 8-pin DIP devices, made by omitting the audio preamp/driver from the 7930.

Figure 2-1 is a block diagram of a 7910 and Fig. 2-2 is a circuit that can be used to drive the IC. The built-in oscillator—two series-connected inverters connected between pins 1, 2, and 3 (terminals OS_1, OS_2, OS_3) and tuned by R_O1, R_O2, and C_O—generates a nominal frequency of 47.5 kHz. That signal is processed by the various internal circuits to develop the notes, tempo, keying, shaping, and other controls needed to produce a specific melody, chime, or alarm.

Pin 4 (terminal MI) is the on-off control point. The tune starts at the beginning of the melody when that terminal is switched HIGH (connected to V_{DD}) and continues until the circuit to terminal MI is opened (switched to LOW). If the terminal is still switched HIGH when the tune ends, the tune is simply played over.

Control terminals CTL_1 and CTL_2 (pins 10 and 11), when connected to R1-C1 and R2-C2, regulate the time constant of the envelope. The overall tone can be changed by varying C1, C2, R1, and R2. Terminal CTL_3 (pin 12) is the unamplified output of the melody-generator section of the IC. Terminal CTL_4 (pin 13) is the input to the built-in audio preamplifier. In Fig. 2, R4 is the volume control. Terminals OUT_1 and OUT_2 (pins 14 and 15) are the preamp outputs used to drive the complementary bipolar output-transistors.

Terminals SEL_1, SEL_2, and MSL (pins 5, 6, and 9) are used to control the output of the IC. The terminals are connected to switches (S1-S3) that are used to select melodies, chimes, or alarm, depending on their setting. Table 2-2 shows the switch settings for a 7910 programmed for two melodies and two alarm sounds. In the MELODY TEST modes, the tempo of the selected melody is stepped up eight times. The switch settings for 7910s that have

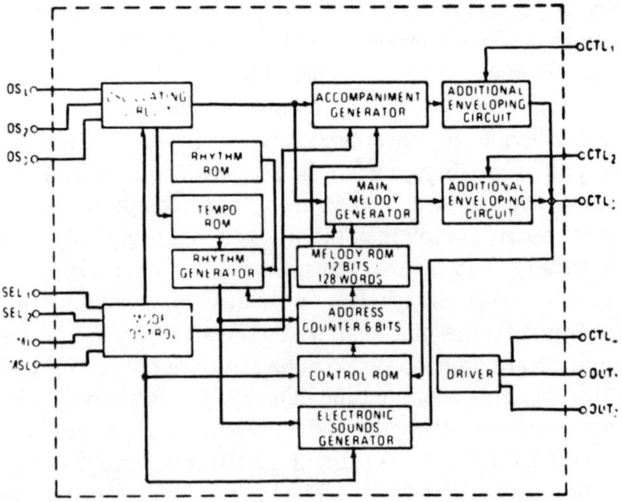

Fig. 2-1.

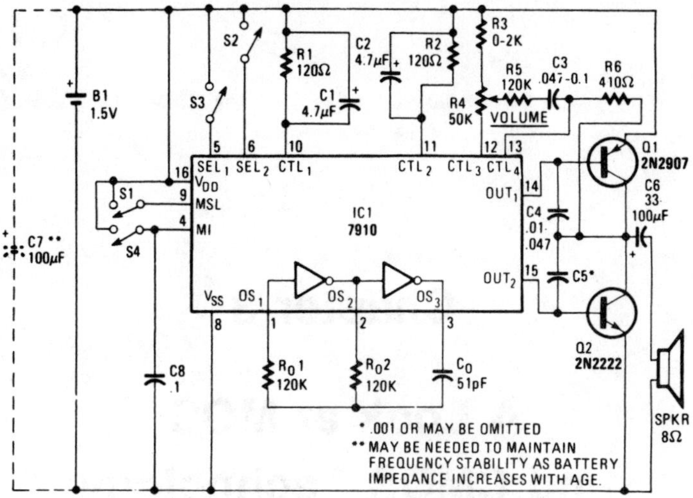

Fig. 2-2.

two melodies and no alarm, or no melodies and two alarm sounds, are similar to those in the table.

I've mentioned the possibility of using a melody generator as a replacement for a doorbell. For that application, I suggest the 7910-O, P, X, or CH versions. Those are different 16-note arrangements of Westminster chimes; the devices are suitable for use as doorbells since they are programmed to play through just once and then stop.

At present, there are fifteen different ICs in the 7910 series, seven in the 7930 series, and two in the 7920 series. The melodies available are too numerous to list here, but, to give you some idea of what is available, they range from "Jingle Bells" and "Silent Night" in the 7910-CU to a Chopin nocturne and a Mozart minuet in the 7910-CE.

Table 2-1.

FUNCTION	S1	S2	S3	S4
Melody 1	on	off	off	on
Melody 2	on	off	off	on
Buzzer	—	on	off	on
Chime	—	off	on	on
Test mode/melody 1	on	on	on	on
Test model/melody 2	on	on	on	on

Chapter 3

A Look at MOS and CMOS Technology

The past two or three years have given us many advances in the field of power semiconductors. Among those developments are increases in operating voltage, operating frequency, and switching speed. The advances in semiconductor technology were needed to keep up with those in power control, switching-type power supplies, Class-D switching-type hi-fi power amplifiers, and other applications requiring fast, efficient transistors capable of handling high voltages and current.

It all began with the MOS (*M*etal-*O*xide *S*emiconductor) technology that gave birth to the MOSFET (MOS *F*ield-*E*ffect *T*ransistor). In the MOSFET, a metal-gate electrode is placed on a silicon dioxide insulator as shown in Fig. 3-1. The drain and source are carrier-doped $p+$ regions in the n-type silicon substrate. Normally, that type of MOSFET acts as an open circuit between the source and drain until a negative voltage is applied to the gate. When that happens, a conducting channel of positively charged holes permits current to flow between the source and drain. (Holes are the absence of electrons. The current carried between source and drain is equivalent to conduction by positive charges; thus p-type conduction through a p channel.) Hole conduction is enhanced (increases) as V_{GS} is made more negative.

An FET of the opposite polarity (called n-channel—for n-type conduction) can be made so current is conducted between source and drain by electrons when a positive voltage is applied to the gate.

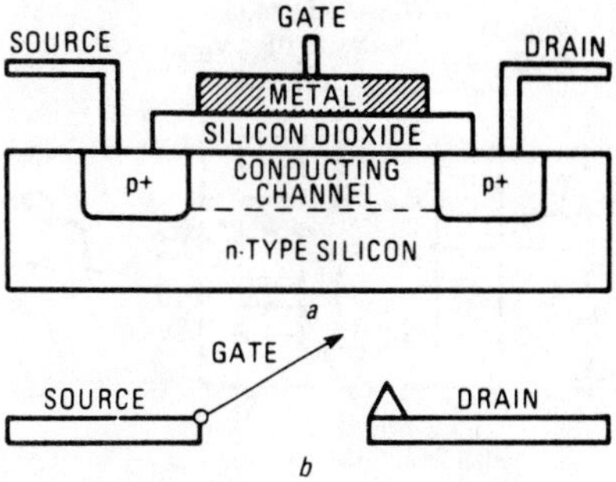

Fig. 3-1.

Both *n*- and *p*- type FETs can be fabricated on the same chip to form a CMOS (*C*omplementary MOS) or COS/MOS (*CO*mplementary-*S*ymmetry MOS—an RCA designation) structure.

CMOS devices are most widely used as logic elements with *p*-type and *n*-type MOSFETs connected in series on the same chip as shown in Fig. 3-2. The devices are normally fabricated on *n*-type silicon material that serves as the substrate for all of the *p*-channel MOSFETs *p*+ channels, diodes, and resistors. A lightly doped *p* well is diffused into the *n*-type base material. Guard bands surround the *n*-channel and *p*-channel MOS devices to prevent leakage between them. The *p*+ guard band on the *n*-channel device serves as the internal ground or V_{SS} bus. The *n*+ guard band serves as a bus connecting the V_{DD} supply to all *p*-channel devices.

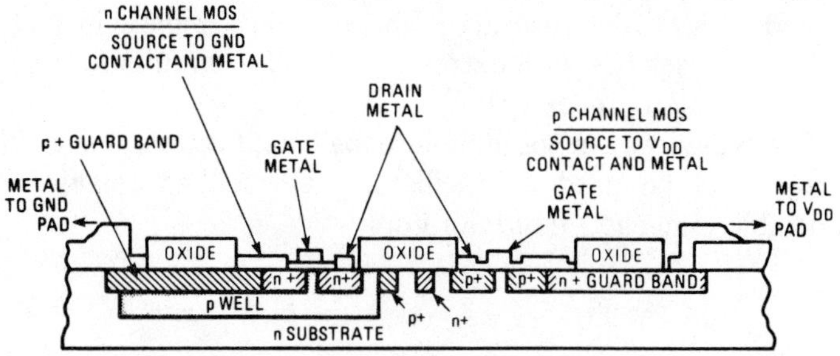

Fig. 3-2.

Figure 3-3 shows an inverted made using an *n*-channel and *p*-channel enhancement-mode FET. The gates are connected in parallel. When a voltage of given polarity is applied to the gates, only *one* transistor can conduct; both cannot conduct at the same time. The substrate of the *p*-channel transistor is at V_{DD} and the substrate for the *n*-channel is at V_{SS}, or ground potential.

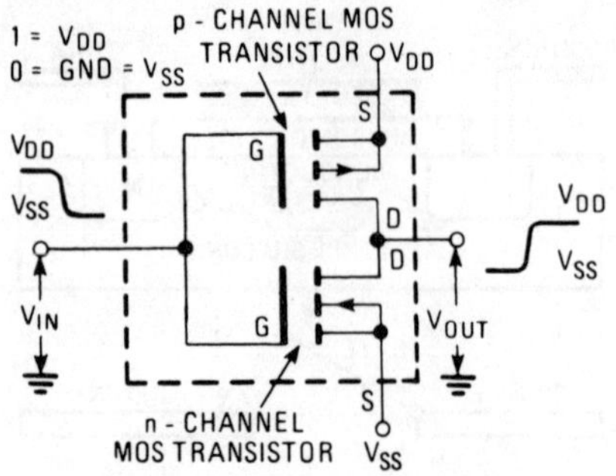

Fig. 3-3.

If the inverter gates are HIGH (logic "1"), the *n*-channel MOSFET is turned on and the *p*-channel MOSFET is turned off. Thus the output is LOW (logic "0"). If the input to the gates is LOW, the FETs change states; the *n*-channel device turns on and the *p*-channel device turns off. Since the CMOS device can be compared to two series-connected normally-open switches, only one of which can be closed at a time, no current can flow from the supply ground.

The quiescent power consumed by the CMOS device is extremely low—equal to the product of supply-voltage V_{DD} and leakage-current I_L. The device draws appreciable current only during the transition from logic "0" to logic "1", and vice versa. Therefore, the average current, and thus the power consumed by the device, is directly proportional to the switching.

The extremely low power consumption, and the fact that thousands of MOS devices can be fabricated on a single chip, makes possible low-cost ICs with extremely complex circuits for use in calculators, microprocessors, etc.

CMOS devices are not limited to logic applications. Gates and inverters can be used as oscillators, as well as astable and monostable (one-shot) multivibrators.

Chapter 4

One IC Voltage Conversion

While experimenting with solid-state circuits, have you ever wished that you could power analog circuits from the supply you are using for digital circuits? Well, it's not impossible. In fact, it's simple when you use the Intersil CMOS ICL7660 voltage converter connected as shown in Fig. 4-1. Apply a positive voltage between pins 8 and 3 and you'll get a negative voltage of the same value between pins 5 and 3. Both voltages are measured with respect to pin 3, which is ground or the negative terminal of the battery or driving power supply.

Thus, in the 7660, we have a unique device that converts a positive-input voltage to a negative-output voltage with a conversion efficiency of 98% when $R_L = \infty$. Power-conversion efficiency if 98% when R_L is 5000 ohms. Output current is greater than 40 mA into a 55-ohm load.

If you need a higher supply voltage for a portion of your circuitry, simply cascade two or more 7660s. Need more current? Connect two 7660s in parallel as shown in Fig. 4-2. That circuit is useful when the load on a single 7660 causes an excessive voltage drop.

Normally, simple power supplies cannot be paralleled with any degree of efficiency because the output voltages and internal impedances of the individual sources are never precisely equal. Thus, when two such supplies are paralleled, the one with the higher output will carry most of the load. In Fig. 4-2, each device

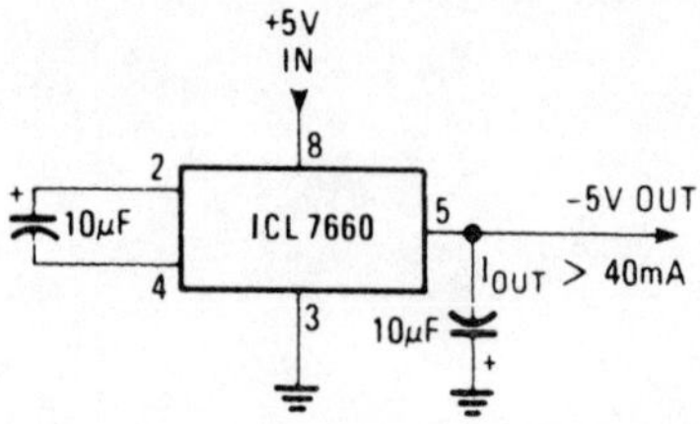

Fig. 4-1.

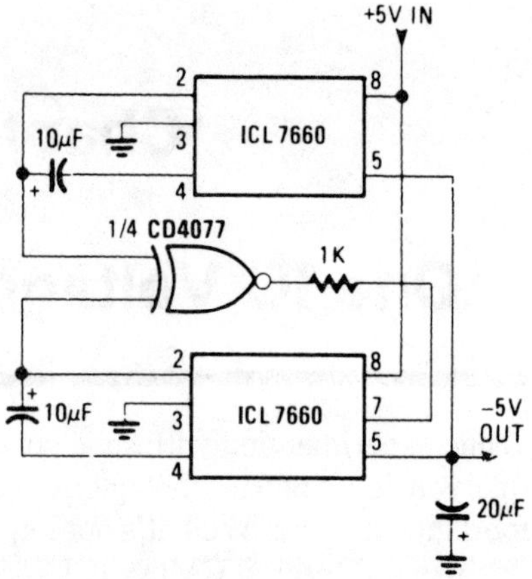

Fig. 4-2.

sees the same input voltage and, since conversion efficiency is nearly ideal, the devices share the load equally.

Each device has a separate "pump" capacitor (10 μF in this application). The CD4077 exclusive NOR gate compares the device outputs at pin 2 and clocks one to maintain sync with the oscillator in the other. The scheme can be extended to deliver around 160 mA by paralleling four devices and using an extra logic gate.

In an interesting circuit innovation described by Intersil, a 1.5-volt cell and a pair of 7660's were used as cascaded voltage doublers to deliver 6-volts DC to a 7126 3½-digit micropower analog/digital converter driving a liquid-crystal display. The circuit is shown in Fig. 4-3.

With 1.5 volts applied to the input of IC1, the first voltage converter, a negative 1.5 volts is developed at pin 5. Converter IC2 sees a total of 3 volts at its input (1.5 volts from the battery and 1.5 volts from IC1) and produces −3 volts with respect to the output of IC1. The total voltage across IC2 pins 8 and 5 is four times the battery voltage, or 6 volts. That is high enough to supply the 7176 A/D converter and the 8069 voltage reference. The external voltage reference is needed because the A/D converter's internal reference works best with a power supply of over 6.5 volts. The 1.2-volt external reference insures that the A/D converter will work correctly even when the battery voltage is low.

Total battery drain is typically only 750 μA. Battery voltages up to 3.5 may be used. Diode D1 should be used whenever battery voltage can be expected to exceed 3 volts.

The ICL7660 is an inexpensive solution to many voltage-conversion problems. It's only $1.95 in lots of 100. Your Intersil distributor or favorite mail-order supply house should have it in stock.

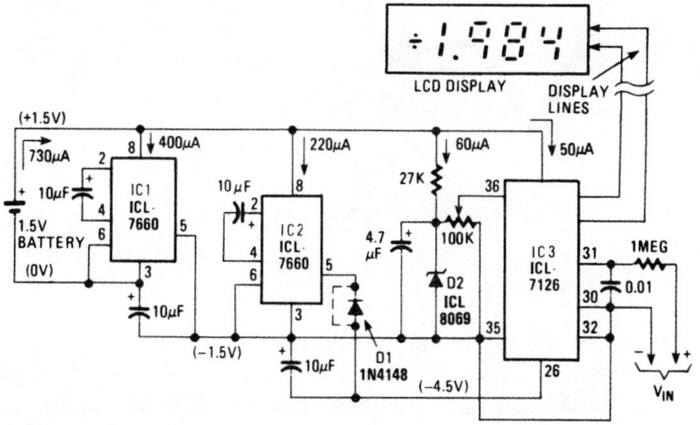

Fig. 4-3.

VMOS POWER DEVICES

Field-effect transistors (FETs) have been around as practical devices since about 1952. But for the next twenty years or so, they were strictly low-power devices capable of handling only loads of less than 1 watt. Thus, they were not able to compete with bipolar transistors and SCRs in power-handling applications. The reason for that is that in a typical FET, the current travels horizontally, just below the surface of the chip, and the maximum current density is much lower than that of a bipolar device that uses vertical current flow.

Around 5 years ago, a new FET technology was developed to increase the current density of the devices. That in turn, led to the development of high-power FETs that feature both high-voltage and high-current capability. The technological improvement on the basic MOSFET, Vertical MOS or VMOS (also called V-groove MOS) owes its greatly improved characteristics and high-power capability to a "V" groove channel that promotes vertical current flow.

Figure 4-4 shows a cross section of a VMOS channel. The device is fabricated on an n+ substrate that becomes the drain and provides a low-resistance current path. An n– epitaxial layer is added to increase the drain-to-source breakdown voltage and to reduce feed-back capacitance.

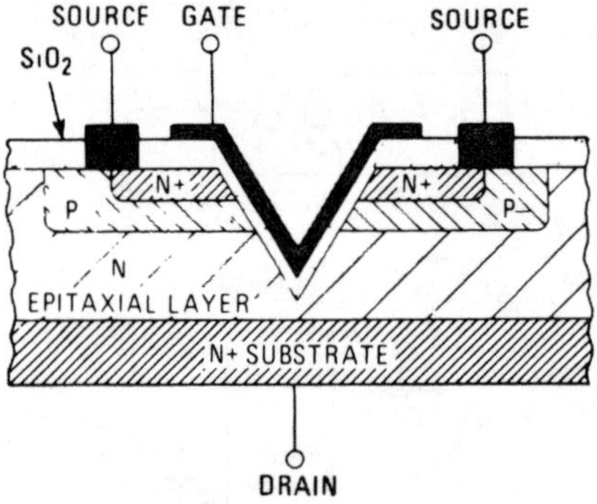

Fig. 4-4.

Lightly doped p– bodies are diffused into the epitaxial layer to form channel regions. Smaller n+ regions are diffused into the p– island to form source regions. The n– epitaxial layer and substrate now become drain regions.

The V-shaped groove is etched through the source and channel regions into the epitaxial layer. A silicon oxide layer is then grown on the surface and in the groove and then aluminum metallization is added to form source and gate connections. Finally, the entire chip is passivated (coated with an inert protective material) to keep contaminants from entering the gate material.

The vertical design of the VMOS technology gives it the following advantages over the conventional MOS structure:

1. The length of the current channel is determined by the depth of the diffusion—a much more easily controlled factor than mask spacings used to define channel lengths in conventional MOS. That makes it possible to obtain a better width/length ratio and allows higher current densities to be obtained.
2. Each groove creates two parallel channels so the current density is inherently doubled.
3. The substrate forms the drain contact so that there is no need

to provide space for a metal drain contact on top of the chip. That reduces chip area and thus keeps saturation resistance low.

4. The high-current density results in low gate-to-drain (feedback) capacitance because the portion of the gate nearest the drain (called the overlap) is in the bottom of the V groove. In a conventional MOSFET, extra drain-gate overlap must be provided to allow for possible mask misalignment. That increases the gate-to-drain and gate-to-source capacitances.
5. The epitaxial layer of the VFET is lightly doped so it absorbs the depletion area from the reverse-biased drain-to-body PN diode. That greatly increases the break-down voltage while having little or no effect on other device parameters.

Various designs of VMOS transistors have been developed by different manufacturers to increase such parameters as speed, power-handling capacity, and voltage and current ratings. Names given those technologies include DMOS, VDMOS, TMOS, ZMOS, HEXFET and SuperFet.

Chapter 5

Advances in IC-Timer Technology

The 555 integrated-circuit timer is probably the most popular and widely used IC since the development of the 709 and 741 op amps. Numerous applications for this device have appeared in nearly all contemporary electronics magazines and in several books devoted entirely to the 555. The device can be used in the monostable mode as a timer, or as a free-running multivibrator when operated in the astable mode.

Exar Integrated Systems has introduced the more sophisticated XR-2243 Micropower Long-Range Timer; a device that can easily surpass the 555-family of ICs in versatility, and popularity among electronics hobbyists and experimenters.

That device is a monolithic timer/controller capable of producing time delays ranging from microseconds to days. The functional diagram for the XR-2243 is shown in Fig. 5-1 and the basic schematic diagram is shown in Fig. 5-2.

Applications for the XR-2243 include: sequential timing, generation of long time-delay periods, precision timing, and ultra-low-frequency oscillation. Its current drain is extremely low—less than 100 μA on standby and less than 1 mA during normal operation. Those minimal current requirements make the XR-2243 highly applicable for use in battery-powered circuits. The XR-2243's basic electrical characteristics are listed in Table 5-1.

Fig. 5-1.

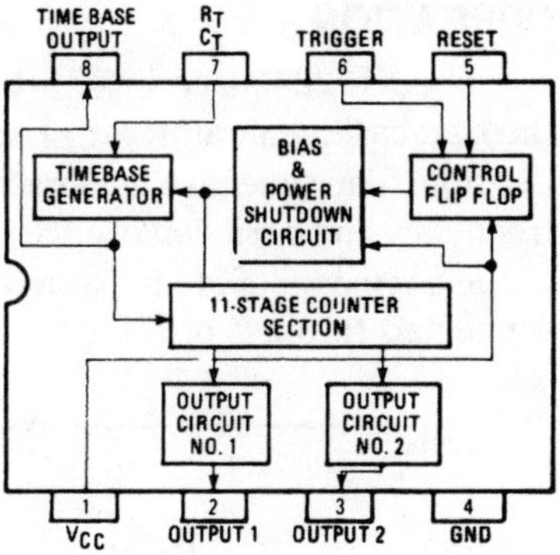

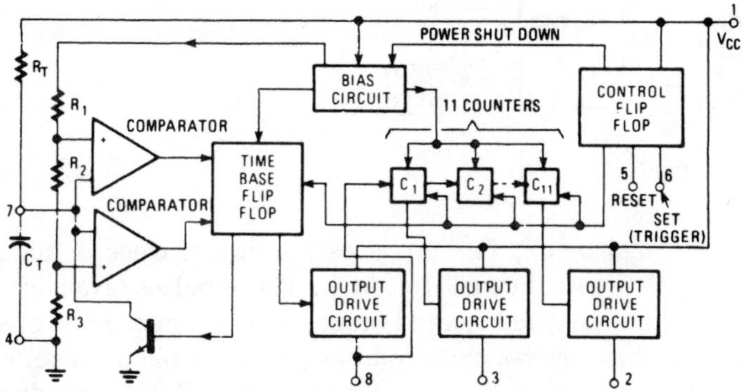

Fig. 5-2.

Table 5-1.

Parameters	Typical values	Conditions
Supply voltage	2.7V min, 15V max	
Standby current	45—80—250 μA	V_{CC} = 2.7*—5—15 V
Operating current	750—900—1250 μA	V_{CC} = 2.7**—5—15V
Timebase timing accuracy	0.5%	V_{CC} = 2.7V**
Temperature drift	80—150—300 ppm/°C	V_{CC} = 5—15—8V
Maximum frequency	35 kHz	
Timing resistor R_T	5K to 10 meg	
Timing capacitor C_T	.005 μF to 1000 μF	
Trigger threshold	1.4V	
Trigger current	22 μA	V_{RS} = 0V, V_{TR} = 2V
Trigger impedance	25K	
Reset threshold	1.4V	
Reset current	22 μA	V_{TR} = 0V, V_{RS} = 2V
Reset impedance	25K	
Power dissipation	300 mW (plastic package)	

*V_{TR} = 0V, V_{RS} = 5V
**V_{TR} = 5V, V_{RS} = 0V

OPERATION

The XR-2243 (see Figs. 5-1 and 5-2) has three main sections: the timebase generator, an 11-stage binary counter and a control flip-flop. The timebase generator is a relaxation oscillator, whose frequency and period of oscillation are determinated by the values of the components of an external R-C network (R_T and C_T) that is connected to pin 7 of the device, as shown in Fig. 5-3.

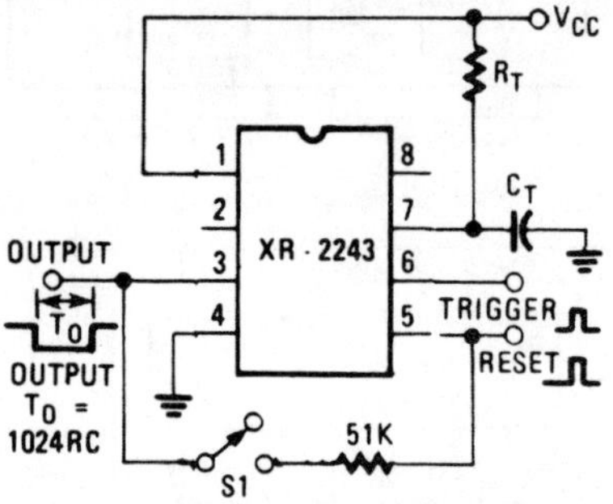

Fig. 5-3.

The timebase flip-flop produces a string of clock or timing pulses with a period equal to R × C. Those pulses (available at pin 8) are internally fed to an 11-stage binary counter. For a given R × C product, the counter produces (at output pin 3) an output-delay pulse of 1024 × R × C. Cascading two XR-2243s produces a total time delay of 1024^2 × R × C or 1,048,576 × R × C. Similarly, cascading three stages will produce a 1024^3 × R × C time delay. Thus, we see that by simply cascading timer stages, we can achieve delays of days, weeks, months, and even years. For example, a controller for time-lapse photography applications can be designed around just a single XR-2243 timer.

The first and last counter stages drive output transistors that are capable of sinking 10 milliamperes. The second output pin (pin 2) delivers a squarewave with a period of 2 × R × C during the timing interval.

The third section of the timer IC is the control flip-flop. That circuit resets each counter stage in the XR-2243 to a logic-1 level and starts the timebase generator when a positive-going trigger pulse is applied to the SET (TRIGGER) pin. That section also activates a shutdown circuit when the timing cycle is completed, or when a positive-going reset pulse is applied to pin 5. When power is shut off, bias voltages are removed from the timebase generator and the counters. Standby power drops to the microwatt range with a typical supply current around 85 μA when V_{CC} is 5 volts.

Figure 5-3 shows the two basic modes of operation for the XR-2243. In the monostable mode—as used for time-delay and timing applications—the output terminal (pin 3) is connected to the RESET pin (pin 5) through an external resistor. The circuit is triggered by a positive-going pulse applied to pin 6. The last counter output goes low as the first timing pulse is fed through from the timebase generator. The pin-2 output remains low until after the 1024th timing pulse has passed.

After the 1024th timing pulse, the output goes high, delivering a positive-going pulse to pin 5. That positive-going pulse terminates the timing cycle and resets the circuit. The time interval between the set and reset actions is $1024 \times R \times C$. During the timing interval, the timebase generator produces, at pin 2, a squarewave output with a period of $2 \times R \times C$.

In the astable mode, the output at pin 3 is not connected back to the RESET pin. A trigger pulse on pin 6 starts the timer. That timer continues to operate in the free-running mode and produces, at pin 8, a squarewave with a frequency that is 1/2048th of the timebase-generator frequency.

Chapter 6

An Audio Preamp and Graphic Equalizer

Figure 6-1 shows how the 5532 dual low-noise high-performance op-amp from either Signetics or Exar can be used in a phono preamp and graphic equalizer. The circuit itself is not very new, but its performance is greatly improved by the 5532 because it develops less noise, has improved output-drive capability, and has greater small-signal and power bandwidths than most common op amps. The characteristics of the 5532 op amp are listed in Table 6-1. The device is internally compensated for unity gain. If you use it where very low noise is of prime importance, you'll want to specify the 5532A version with its guaranteed noise specifications.

Returning to the schematic, the phono-preamp stage, IC1-a, provides standard RIAA equalization for a magnetic cartridge. The necessary bass-boosted response is provided by the R-C network in the op amp's feedback loop. The cartridge loading-resistor, R1, has a value of 47K to match the impedance of the average magnetic cartridge. Adjust R1's value as required if your cartridge works best into a different load resistance.

The graphic-equalizer section consists of an input buffer (IC1-b), variable-boost/cut active filter (IC2-a), and output summing-amplifier (IC2-b). The IC1-b circuit is designed for unity gain and is used mainly for impedance-matching between the preamp and the equalizer (active filter) input. The filter is a variable-bandpass or notching device, depending on the setting of control R9.

Table 6-1.

- Small-signal bandwidth: 10 MHz
- Output drive capability: 600Ω, 10V (rms)
- Input noise voltage: 5nV/√Hz
- DC voltage gain: 50000
- AC voltage gain: 2200 at 10 kHz
- Power bandwidth: 140 kHz
- Slew rate: 9V μs
- Large supply voltage range: ± 3V to ± 20V

Any number of equalizer filter-stages can be used within the range of about 20 Hz to 20 kHz. However, the more stages you have, the easier it is to boost or cut a particular frequency without affecting the response at adjacent frequencies. All the filter stages use the same R-C feedback-network configuration shown in Fig. 6-1, to provide a maximum of about 15-dB of boost or cut at f_0, the center frequency. The only differences in each stage are in the values of C5 and C6, which set the values of f_0. Table 6-2 lists the values for C5 and C6 for 22 center frequencies in the audio spectrum. Note that C5 is ten times as large as C6 and that the values for R8 and R10 are both related to the value of R9 by about a factor of 10. The center frequencies have been adjusted so that C5 and C6 are standard, off-the-shelf, values. We recommend using linear slide-potentiometers for R9.

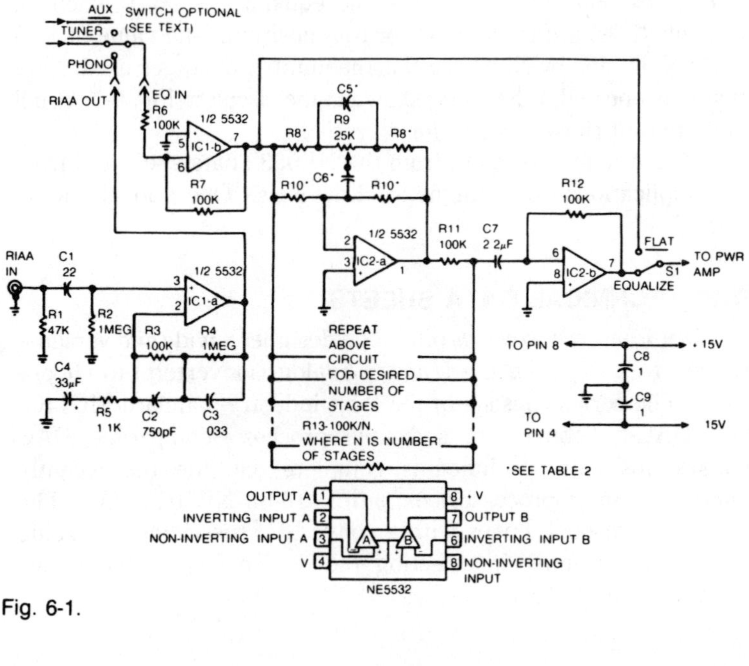

Fig. 6-1.

R8 2.4K R9 25k R10 240k

f_o	C5	C6
23 Hz	1μF	.1μF
50 Hz	.47μF	.047μF
72 Hz	.33μF	.033μF
108 Hz	.22μF	.022μF
158 Hz	.15μF	.015μF
238 Hz	.1μF	.01μF
290 Hz	.082μF	.0082μF
350 Hz	.068μF	.0068μF
425 Hz	.056μF	.0056μF
506 Hz	.047μF	.0047μF
721 Hz	.033μF	.0033μF
1082 Hz	.022μF	.0022μF
1588 Hz	.015μF	0015μF
2382 Hz	.01μF	.001μF
2904 Hz	.0082μF	820pF
3502 Hz	.0068μF	680pF
4253 Hz	.0056μF	560pF
5068 Hz	.0047μF	470pF
7218 Hz	.0033μF	330pF
10827 Hz	.0022μF	220pF
15880 Hz	.0015μF	150pF
23820 Hz	.001μF	100pF

Table 6-2.

The value of R13 depends on the number of filter stages used. It ensures that the gain across the equalizer is unity when all controls (R9s) are in the FLAT or 0 dB position. The value of R13 is 100K divided by N, where N is the number of stages used. Note that only one audio channel is shown in the circuit in Fig. 6-1. You'll need two of those circuits for stereo.

This circuit was taken from the NE5532 data sheet and from the applications note "Signetics Low-Noise Operational Amplifiers."

DAC TECHNICAL DATA SHEETS

D/A Converter Products offers the designer a wide and versatile range of eighteen DAC's (*D*igital-*A*nalog *C*onverters) to choose from. The series consists of the 8-bit industry-standard MC1408 and DAC-08 converters and microprocessor-compatible DAC subsystems. Also included is complete data on the recently announced microprocessor-compatible 10-bit NE5020 DAC. The 67-page data book opens with a converter cross-reference guide to aid the designer in selecting the device best suited to his requirements.

Chapter 7

DC Voltage-Controlled Switches

In this chapter we'll take a look at a DC-controlled switch that can replace the usual rotary or pushbutton used to connect various signal sources to the input of an audio amplifier.

National Semiconductor has just announced the LM1037 and LM1038 DC Controlled Audio Switches. Both are designed to allow any one of four stereo-input sources to be connected to the stereo-output lines. Channels are selected by applying a DC voltage level to the channel-select control pins. Features of the LM1037/38 switches are:

- Wide supply-voltage range (5 to 30 volts)
- Click-free operation
- Low distortion, 0.04% typical
- High signal-to-noise ratio
- High audio-input impedance
- Low audio-output impedance
- Wide control-voltage range (2 to 50 volts)
- High input-impedance on select pin

The LM1037 has four control or channel-select pins; one for each stereo-input source. If no channel is selected, the outputs are muted and clamped to the DC output level.

The LM1038 is designed so that channels can be selected by BCD input pulses. Two input pins feed clock-enabled latches so

Table 7-1.

Parameter	Min	Typ	Max	Units	Conditions
Supply voltage	5	12	30	volts	
Supply current (LM1037)		7		mA	V_{supply} = 12
Supply current (LM1037)		10		mA	V_{supply} = 30
Supply current (LM1038)		14		mA	V_{supply} = 12
Supply current (LM1038)		20		mA	V_{supply} = 30
Voltage gain	0.2	0		dB	
Signal handling			3 2	$volts_{rms}$	V_{supply} = 12
Distortion (THD)		0 04	0 1	percent	1 V_{rms} input @ 1 kHz
S/N Ratio (CCIR)		100		dB	1 V_{rms} input @ 1 kHz
Crosstalk between stereo channels	80	100		dB	1 V_{rms} @ 1 kHz
Crosstalk between connected and unconnected channels	80	100		dB	1 V_{rms} @ 1 kHz
Relative output in muted state	80	90		dB	1 V_{rms} @ 1 kHz

that the BCD pulses can be strobed from a bus. Or, the clock-input pin can be permanently enabled and the channels selected by DC voltage levels. A separate pin allows the output to be muted, via either a pulse or a DC level.

As of this writing, only preliminary specifications (Table 7-1) on the LM1037/38 have been released, so we don't have complete details on the structure of the semiconductor device used as the switch elements; but I'll bet that the switch is some form of FET. In similar analog switches from other manufactures, JFETs, MOSFETs and CMOS (N-channel and P-channel FETs in parallel) have been used as gates or switches. Figure 7-1(a) is a functional diagram of a stereo-audio switch. When the control voltage is applied, the switches close so that the input signals pass through to the output terminals. Figure 7-1(b) is the circuit of a basic analog switch using a JFET as the switch element.

Figure 7-2 is the functional block diagram of the LM1037. The channel-selection details are outlined in Table 7-2.

UNIQUE MONOLITHIC ACTIVE FILTER

National Semiconductor's MF10 is the first monolithic active filter that does not require external capacitors to operate, and is

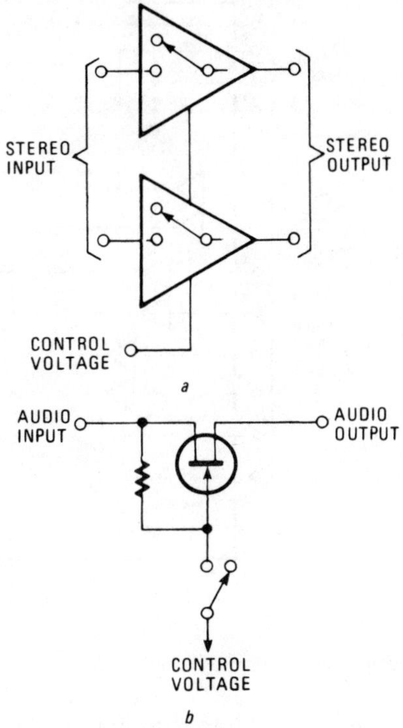

Fig. 7-1.

capable of performing a variety of functions. It is ideal for signal processing, telecommunications, modems, instrumentation, and audio equipment. The center frequencies of all previously available active filters are adjusted with external resistors and/or capacitors.

Table 7-2.

Inputs switched to output pins

Input channel		1A	2A	1B	2B	1C	2C	1D	2D	
Pin no.		2	4	6	8	11	13	17	15	12
DC control	1	VL	VL	VL	VL	VH	VH	VL	VL	VL
conditions	3	VL	VL	VL	VL	VL	VL	VH	VH	VL
(pins)	16	VH	VH	VL	VL	VL	VL	VL	VL	VL
	18	VL	VL	VH	VH	VL	VL	VL	VL	VL
Output pin		10	9	10	9	10	9	10	9	9 & 10 (MUTE)
Output channel		1	2	1	2	1	2	1	2	

Low switching-level: VL · 0.8 volts
High switching-level: VH · 2.0 volts

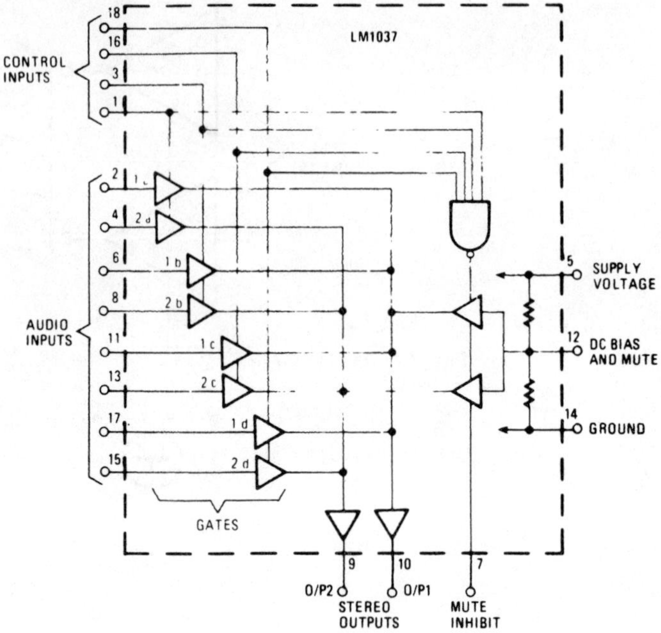

Fig. 7-2.

The MF10 filter is unique in that the center frequencies of its various second-order functions are directly proportional to an external clock frequency and fixed at an accuracy of 0.6%, or directly proportional to both the clock frequency and external resistor ratios. Gain and selectivity (Q) are adjusted using external resistors.

The clock and three or four resistors are the only external components necessary to operate the MF10 so the device is less sensitive to external component variations. Frequency stability is directly dependent on the quality of the clock. A single clock can be used to drive an almost infinite number of MF10 filters.

Unlike other available monolithic filters which have been designed for a specific function, the general-purpose MF10 can perform a wide variety of functions including allpass, highpass, bandpass, and notch. Up to fourth-order functions, and any configurations such as Butterworth, Bessel, Cauer, and Chebyshev can be had by cascading the two second-order building blocks. The center frequency can be as high as 20 kHz with a Q of 500. Typically, the lowpass and bandpass outputs can sink 0.75 mA and source 3 mA while the notch, allpass, and highpass outputs sink 1.5 mA and source 3 mA.

OPTOCOUPLERS FEATURE BACK-TO-BACK LED'S

Motorola has introduced two new dual-LED Optocouplers designed specifically for AC input applications. The devices—the HIAA1 and HIAA2—consist of two gallium-arsenide (GaAs) infrared-emitting diodes connected in inverse parallel, and optically coupled to a single silicon phototransistor (detector) in the standard 6-pin DIP package.

The Optocouplers can be used to detect the presence, absence, or interruption of AC power and trigger a desired action. A major application is the detection of a ring signal on telephone lines while providing isolation from the ring lines. The HIAA1 and HIAA2 are directly interchangeable with GE devices bearing the same type numbers, while providing for up to 7500 volts.

TUNABLE LOW-PASS SAMPLED-DATA FILTERS

The MC145414 dual-tunable low-pass filter from the Motorola MOS Integrated Circuits Group is the latest addition to the growing family of CMOS switched-capacitor filters.

The MC145414 has a typical power dissipation of 30 mW when active and 1 mW in the power-down condition. Operating voltage can be obtained either from a single or split power-supply with a voltage range of 10 to 16 volts. It can interface to either TTL or CMOS devices. The device is useful in LPC (*L*inear *P*redictive *C*oding) or CVSD (*C*ontinuously *V*ariable *S*lope *D*etection) speech applications, modems, or other applications where a precision low-pass filter is needed. Specifications include a dynamic range better than 80 dB, passband ripple less than 0.3 dB, and idle noise typically 100 dBrnCO (Decible RMS noise C-message weighted referenced to zero).

Chapter 8

A New High-Power Op-Amp

I have wondered, and I suppose you have also, about the performance and the circuitry involved in those LSI audio-power amplifiers that are offered by a number of mail-order electronics supply houses. I haven't been able to come up with any technical data on those devices but was fortunate in running across an application note on a new and interesting device from National Semiconductor. It is the LH0101 low-distortion high-power wideband operational amplifier designed to deliver a high current into a variety of loads. It is conservatively rated at 2 amps with negligible crossover (zero-crossing) distortion. Frequency response is from DC to above 4 MHz. It is in a hermetically sealed TO-3 package. Table 8-1 shows the typical performance characteristics at 25 °C ambient and a +15-volt supply.

The LH0101, shown schematically in Fig. 8-1, has three basic sections: an op amp, buffer amplifier, and power amplifier. The op amp uses a BI-FET configuration to take full advantage of the superior DC performance offered by the FET input and the desirable slew rate, settling time, and low bias-current characteristics of this type of device. In addition, the internal frequency compensation makes the BI-FET an ideal around which to design a power amplifier.

Most power amplifiers designed for high-current output over a wide frequency range are either designed for Class AB or Class B operation. Both of those designs have a tendency to produce

Table 8-1.

Parameter	Conditions	Value
Output current		2A
Input offset voltage		5mV
Input bias current		50pA
Input offset current		25pA
Input resistance		$10^{12}\Omega$
Large signal voltage gain		200V mV
Output voltage swing	$R_L = 100\Omega$	±12.5V
	$R_L = 10\Omega$	±11.6V
	$R_L = 5.0\Omega$	±11V
Slew rate	$A_V = +1$	10V μs
Full power bandwidth	$A_V = +1, R_L = 10\Omega$	300kHz
Small signal rise time	$A_V = +1, R_L = 10\Omega$	100 NS
Small signal setting time to 0.01%	$V_{IN} = 10V, A_V = +1$	2μs
Gain bandwidth		4 MHz
Harmonic distortion	$f = 1kHz, P_O = 1W$	0.005%
	$R_L = 10\Omega, A_V = +1$	
	$f = 20kHz, P_O = 1W$	0.05%
	$R_L = 10\Omega, A_V = +1$	

crossover distortion. For minimum crossover distortion, a power amplifier must maintain a low output impedance throughout zero-crossing. To do that, the push-pull output transistors must smoothly drive the load, alternately switching current-sinking and current-sourcing duties at the crossover point.

In a Class B configuration, both output transistors are completely cut off at the crossover point. Thus, output impedance is relatively high and crossover distortion is severe. In a Class AB design, both output transistors are biased on during no-load conditions, thus providing a low output resistance and thereby eliminating crossover distortion.

However, in a Class AB design, crossover distortion can develop with high-level input signals. For example, when the input-signal voltage causes full output current to be delivered to the load, the increased base-emitter voltage of the driving transistor tends to bias the resting transistor off. Now, when the input signal reverses polarity, so that the output swings negative, the amount of crossover distortion, if any, depends on how fast the resting transistor can turn on and assume its share of duty cycle. The condition worsens as the frequency of the input signal increases.

The output stage of the LH0101 combines both Class B and Class AB designs to achieve smooth distortion-free switching at the crossover point.

The buffer stage of the LH0101 (Fig. 8-1) is a unity-gain current amplifier consisting of transistors Q3-Q11 and Q5 and Q10. Operating in the Class AB mode, what the buffer does is to provide distortion-free drive during the zero crossing. Bandwidth extends beyond 50 MHz to eliminate the possibility of bandwidth-induced distortion.

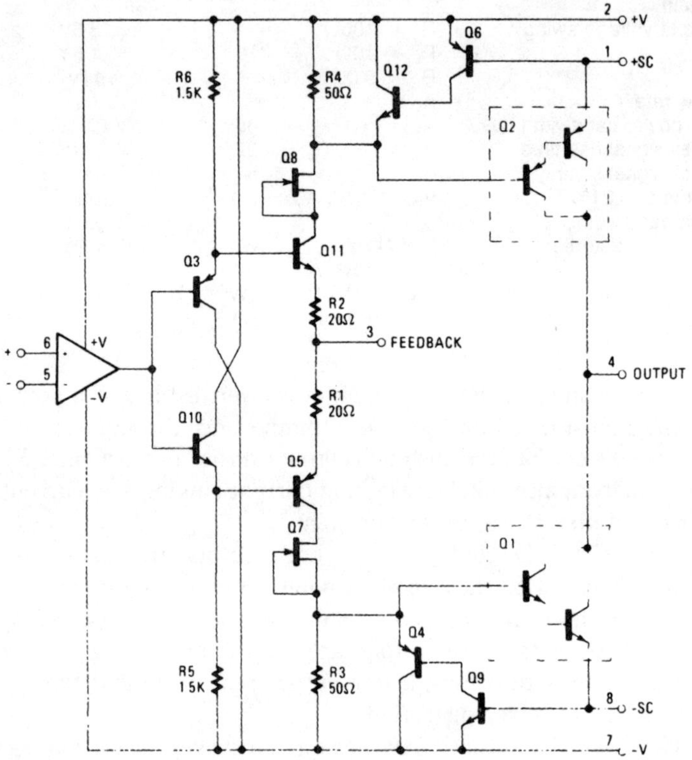

Fig. 8-1.

FETs Q7 and Q8 limit the buffer-stage output current to 50 mA. However, the output stage, consisting of Darlington transistors Q1 and Q2, is set up so that both transistors turn on as the output-load current reaches 25 mA. Under operating conditions, the buffer drives the load at currents up to 25 mA. Above that point, the output stage takes over, delivering power up to the rated output limit. Thus, the power-driving ability of the buffer stage is used to "smooth" the turn-on delay of the output stage and eliminate crossover distortion.

Transistors Q6 and Q9 are in the circuit to prevent the output stage from being over-driven. Current-sensing resistors (R_{SC}) may be connected between the supply and SC terminals to set the limiting level. A drop of approximately 0.6 volt across a sensing resistor turns on either Q6 or Q9. That in turn, turns on Q12 or Q4, respectively, to prevent excess base current from driving the output stage beyond the design limit. Current-sensing resistors $R_{SC} = 0.6/I_{SC}$. When $I_{SC} = 2$ amps, $R_{SC} = 0.3$ ohms.

LOW DISTORTION 40-WATT POWER AMP

Figure 8-2 shows how two LH0101's can be used in a bridge configuration to obtain maximum available power output from a specified supply voltage. Amplifier distortion curves are shown in Fig. 8-3. A slew rate of 10 volts-per-μs extends the full-power bandwidth to beyond 100 kHz.

Application Precautions

In this and other high-current high-power amplifiers, particular attention must be given to ground connections and the length and diameter of PC traces carrying high currents. Keep them short to minimize the development of error voltages. Figure 8-4 shows a suitable method of circuit grounding. The heavy lines represent paths or traces carrying high currents.

The importance of minimizing error voltages can be seen as we examine the current-sensing circuitry in the amplifier in Fig. 8-2. The current-sensing resistors are R1, R2, R3, and R4; 0.15-ohm, 2-watt units that develop the 0.6-volt needed to trigger the current-limiting circuit. A PC trace with a resistance of only 10 milliohm (0.01 ohm) carrying 2 amps will develop a 20-mV error voltage. Add to that the possible error voltages that may develop across the 5-milliohm resistance of a good solder joint and the 10-milliohm resistance of a socket contact.

A heat sink is a *must* to keep the LH0101's operating temperature within a safe range. It should have a thermal resistance of 3.5°C-per-watt ambient. A typical heat sink with that rating, and suitable for a TO-3 device package,is the Thermalloy 6141. It should be mounted with a mica insulator and a liberal application of a thermal-contact fluid or silicone grease.

OTHER APPLICATIONS

The LH0101 is ideally suited for service as a programmable

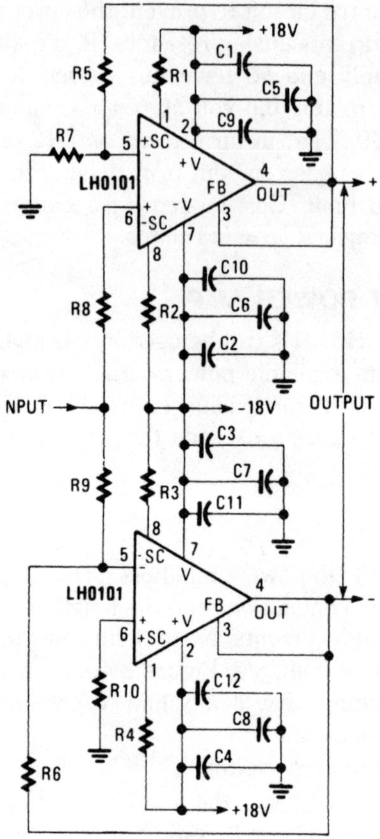

Fig. 8-2.

R1-R4	CURRENT LIMIT RESISTOR	0.15Ω, 2W
R5	FEEDBACK RESISTOR	5K
R6	FEEDBACK RESISTOR	15K
R7-R10	INPUT RESISTORS	10K
C1-C4	BYPASS CAPACITORS	47μF, 25V ELECTROLYTIC
C5-C8	BYPASS CAPACITORS	10μF, 25V TANTALUM
C9-C12	BYPASS CAPACITORS	0.1μF, 25V CERAMIC

current source, coaxial cable driver, CRT yoke driver, and a driver for inductive loads. For information on adapting the device to those applications, refer to Application Note AN-261—*Low-Distortion Wideband Power Op Amp* and *LH0101 Power Operational Amplifier*

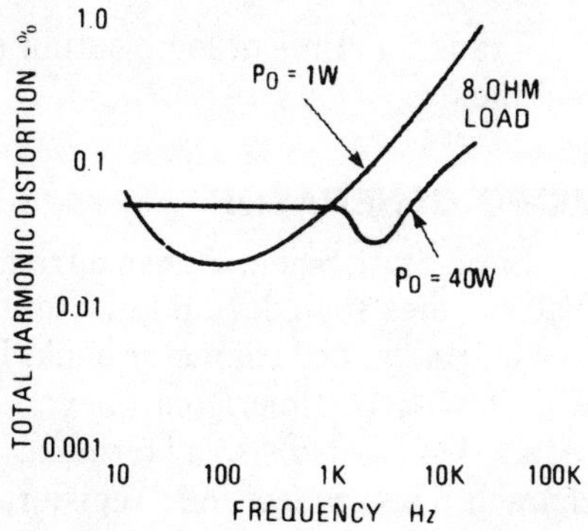

Fig. 8-3.

Data Sheet available from National Semiconductor, 2900 Semiconductor Drive, Santa Clara, CA 95051.

DIVIDE-BY-FOUR-PRESCALER

The RCA CA3199E divide-by-four prescaler takes signals in the VHF/UHF band (up to 1.3 GHz) and reduces them to low-frequency logic levels. The device's high sensitivity eliminates the need for preamplification in most cases. Applications include digital frequency synthesis in VHF/UHF receivers, frequency standards, and as high-frequency dividers in UHF timers and counters.

Accepting either single- or double-ended AC-coupled input signals, the CA3199E provides complementary emitter-follower outputs at standard ECL levels. With unloaded outputs, the typical logic 1 level is 4.2 volts while the logic 0 is 3.4 volts. The device operates from 5 $\pm$0.5 volts. The nominal input signal is a 100 mV sinusoidal waveform in the range of 100 MHz to 1000 MHz; the maximum RMS input voltage is 0.5 volt.

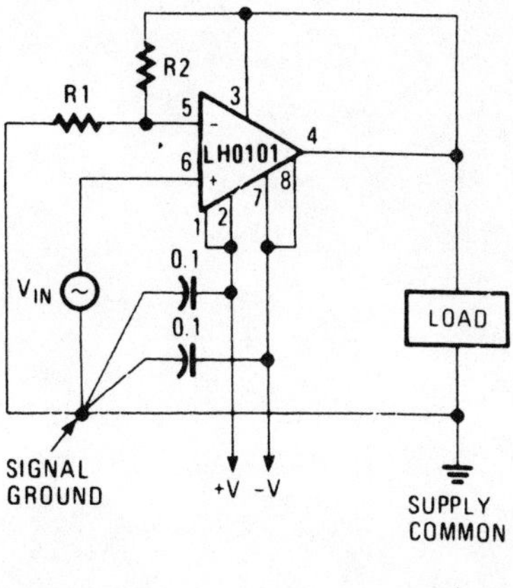

Fig. 8-4.

Transition time of logic output is 0.6 ns for both risetime and falltime.

VIDEO GENERATOR

Solid State Scientific has introduced the SND video generator that includes such desirable features as reverse video, underline, strikethrough, and character blink. It operates with the company's SND5037 CRT timer/controller to provide the display functions required of a video display terminal. The device has an expandable character set, video shift register, four character and graphics modes, and two programmable blink rates.

MICROMINIATURE INFRARED LED

Motorola expands its line of infrared LEDs with the addition of the MLED15 in a tiny clear plastic housing only 0.092 inch in diameter and 0.058 inch high. It has an output of 1.3 mW at a forward peak current of 30 mA. The peak emission wavelength is 930 nanometers so the MLED15 is physically and spectrally matched to the MRD150 phototransistor detector for use in interrupter modules and reflective couplers.

Chapter 9

Measuring Relative Humidity

A knowledge of the relative humidity at any given time can permit us, or others around us, to make decisions that can affect our health, comfort and personal safety. In addition, the measurement of relative humidity (abbreviated RH) is vital in such areas as food processing, air conditioning, packing, photography, paper and lumber production, and chemical manufacturing. Knowing the RH—along with the temperature—can allow airport operators to predict fog and runway icing. Similarly, farmers and nurserymen can predict dew and frost; and highway safety authorities can forecast dangerous fog and icing on bridges.

Electronic circuits for humidity measurement have rarely appeared in the press because humidity-sensitive transducers have been expensive and seldom available to the electronics constructor and hobbyist. Now, for about ten-dollars-worth of semiconductors and discrete components, and less than sixty dollars for the humidity sensor, you can build a direct-reading electronic hygrometer. The device, described in National Semiconductor's Application Note AP 256, is designed around the PCRC-55 humidity sensor, from Phys-Chemical Research, and a linear 0 to 10-volt DC meter.

THE SENSOR

The electro-humidity sensor has a hygrometric element whose impedance changes with changes in relative humidity. The hygrometric portion of the sensor is on the surface of a chemically treated styrene copolymer plastic wafer. Water vapor is sorbed or desorbed by means of adsorption (the adhesion of molecules to a surface; not the same as *ab*sorption). That results in the sensor's having a very rapid response to changes in relative humidity. A single sensor covers the complete range of relative humidity—from 0 to 100%.

The sensor should be excited only by AC voltages (preferably sinusoidal waveforms) of at least 20Hz with no DC component. Sustained operation with a DC voltage, or AC with a DC component, causes a shift in calibration. The maximum allowable current is one mA.

Figure 9-1 is a block diagram of the electronic hygrometer. An amplitude-stabilized squarewave drives a precise alternating current through the sensor. The sensor's output current feeds a current-sensitive logarithmic amplifier to linearize the response. The output of the log amplifier is scaled, rectified, and filtered to provide a 0—10-volt DC-output representing RH's from 0 to 100%.

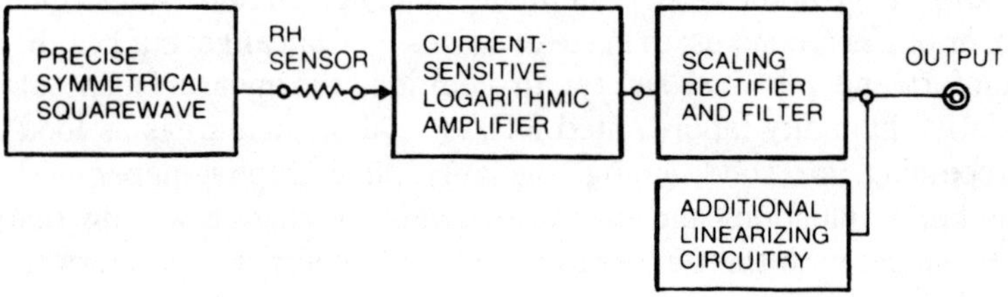

Fig. 9-1.

A PRACTICAL CIRCUIT

In Fig. 9-2, a symmetrical squarewave is generated by IC1-a, an op-amp with positive feedback applied to cause it to oscillate. The combination of constant-current source IC4 and its associated diode-bridge clamps the squarewave output of IC1-a at +8 volts. IC4 has a positive temperature coefficient of 0.033%/°C, which almost completely compensates for the negative 0.036%/°C temperature coefficient of the PCRC-55 sensor. (Mount IC4 close to the sensor so they will be at the same ambient temperature; that way the temperature coefficient of the complete instrument will approach zero.)

The squarewave is buffered, and then fed through the sensor into the summing junction of IC1-c. On negative-going halves of the squarewave, the V_{BE}/I_C characteristic of Q1, in the IC1-c

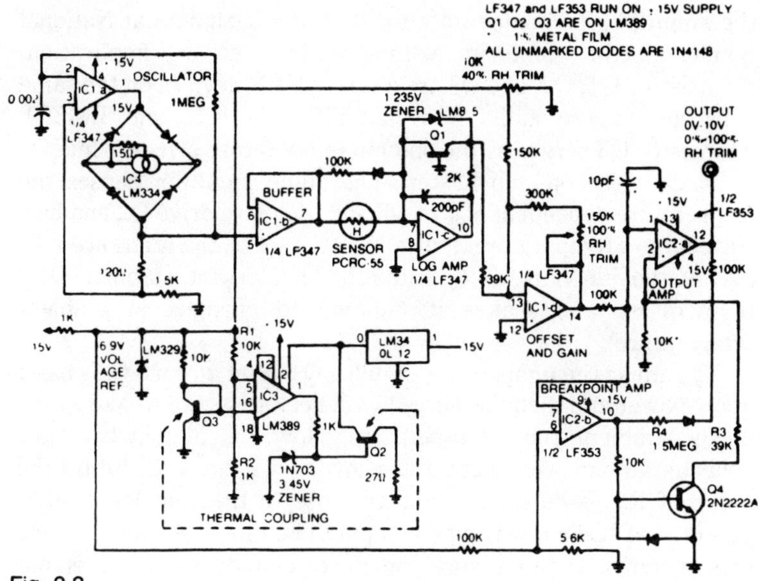

Fig. 9-2.

feedback loop, gives the amplifier a logarithmic amplitude-response. On positive-going half-cycles of the squarewave, feedback through the diode to the summing junction ensures that this point remains at a virtual ground so the sensor always "sees" the required symmetrical drive-waveform.

The output of the log amplifier goes to IC1-d, an op-amp used to sum-in the calibration at the 40% RH point, and to provide adjustable gain for trimming the output to the proper level for a 100% RH reading. The output of IC1-d is filtered to DC by the 100K resistor and 10 μF capacitor and then fed to IC2-a, the output amplifier.

IC2-b compensates for the sensor's departure from linear response below 40% RH. It does that by altering the gain of the output amplifier when its input drops to about 0.36 volt-corresponding to 40% RH. The inverting input of IC2-b, the breakpoint amplifier, is tied to the non-inverting input of the output amplifier. When the input to those paralleled gates drops below 0.36 volt, the output of IC2-b swings positive; that turns on transistor Q4 to produce the required change in the output amplifier. Transistor Q4 is turned off for RH values above 40%, so linearity is then determined solely by the log amplifier.

Transistor Q1, in the log-amplifier feedback lop, is extremely temperature sensitive and can adversely affect the performance of the amplifier. To compensate for that, the designers at National Semiconductor came up with a unique circuit application. Transistors Q1, Q2, and Q3 are discrete NPN devices on the same substrate as the audio power-amplifier in IC3, and LM389. Transistor Q3 serves as the on-chip temperature sensor while Q2 is used as the on-chip heater. The audio amplifier senses the temperature-dependent V_{BE} of Q3 and uses it to drive Q2 and heat the chip to a temperature (typically 50 °C) set by the reference voltage at the junction of R1 and R2. That circuit stabilize Q1's temperature and makes it immune to changes in ambient temperature.

To adjust the temperature-stabilizing circuit, ground Q2's base, apply power and then measure Q3's collector voltage. Make a note of the ambient room temperature, Now, calculate what Q3's collector voltage will be at 50 °C; allowing – 2.2mV/ °C. Adjust the value of R2 to develop a voltage close to the calculated 50 °C potential at IC3's inverting (–) input. The range of values for R2 can be fairly wide because the exact chip temperature is not important as long as it is stable. Finally, remove the ground from Q2's base and the chip should reach the predetermined stable operating temperature within 100 ms. You can check temperature stability by reading Q3's collector voltage while blowing on IC3. The measured voltage should remain constant within 100 μV (0.05 °C).

CALIBRATION

To calibrate the electronic hygrometer, connect a 35K resistor in place of the sensor and adjust the 150K 100% RH TRIM pot for 10 volts output. Now, substitute an 80-megohm resistor for the sensor and set the 40% RH TRIM pot for 4 volts at the output terminal. Repeat the 100% RH TRIM and 40% RH TRIM adjustments until they no longer interact with each other. Finally, substitute a 60-megohm resistor for the sensor and select a 39K resistor for R3; that will develop a 2.4-volt output corresponding to 24% RH. It may be necessary to select a particular 1.5-megohm resistor for R4 to minimize jitter in the meter reading around the 24% RH point.

To obtain the latest pricing and data on the humidity sensor, write to Phys-Chemical Research Corp., 36 W. 20th St., New York, NY 10011. Be sure to specify the PCRC-555 because the company has other sensors whose characteristics are not suitable for the circuit that has been described here.

INSTRUMENTATION AMPLIFIER IC

The standard three-op-amp instrumentation amplifier with its more than a half-dozen discrete external components has been replaced by National's LM363. A low-cost, high-performance monolithic instrumentation amplifier, the LM363 (and the LM163 military version) comes in an 8-pin TO-5 package. It requires no external resistors for accurate fixed gains of either 10, 100, or 500 (one gain per package). The device has a super-beta bipolar input stage that yields a very low input-voltage noise, and a high common-mode rejection ratio (CMRR). For added versatility, the internally set gains can be increased by adding external resistors.

SUPER-FAST OP-AMP

The Harris HA-2539 op-amp features a 600V/μs slew rate and a 600-MHz gainbandwidth product, making it an ideal device for use in pulse- and video-amplifiers, wideband amplifiers, highspeed sample-and-hold circuits, and RF oscillators. Full $\pm$ 10-volt output swing and high open-loop gain make this Harris device useful in high-speed data-acquisition systems.

Power bandwidth is 9.6 MHz, offset voltage is 3 mV, and input-voltage noise is 15 nV/Hz. The device is packaged in a 14-pin ceramic DIP. The HA-2539-2, manufactured for military use, operates in the $-55°$ to $+125°C$ temperature-range. The HA-2539-5, intended for commercial applications, operates in the 0 °C to $+75°C$ temperature-range.

Chapter 10

Digital Lock Circuit

One of the most interesting ICs to come to my attention is the LS7220 digital-lock circuit from LSI Computer Systems, Inc. It is a monolithic PMOS keyless combination-lock designed especially for automotive security applications. It includes sequential logic for reading correct key-closures and detecting out-of-sequence entries. In addition, the circuit includes a "save" input that stores the states of all outputs so that the driver can leave the security system disabled to allow the car to be driven by parking lot and garage attendants. Other features include:

- Stand-alone lock logic
- 5040 possible 4-digit combinations
- Out-of-sequence detection
- Automatic chip enable (for automotive use)
- Programmable convenience time-delay
- Low current consumption (40 μA@ 12 VDC)
- Single power supply (+5 to +18 VDC)

Output voltage specifications are given in Table 10-1. Figure 10-1 shows the pinout of the LS7220, and Fig. 10-2 shows a block diagram of the IC.

USING THE IC

A practical circuit for an automobile anti-theft lock is shown

Table 10-1.

		Source Current			
	VSS	Min	Typ	Max	Units
Lock Control	5 VDC	2.40	3.75	6.30	mA
Output Pin 13	9 VDC	7.20	9.75	14.70	
On (Logic "1")	12 VDC	10.80	14.25	21.00	
$V_{OUT} = V_{SS} - 2$	15 VDC	14.40	18.75	27.30	
	18 VDC	18.00	23.25	30.00*	
Convenience	5 VDC	0.20	0.29	0.50	mA
Delay Pin 12	9 VDC	0.55	0.75	1.13	
On (Logic "1")	12 VDC	0.83	1.10	1.60	
$V_{OUT} = V_{SS} - 2$	15 VDC	1.10	1.44	2.10	
	18 VDC	1.40	1.80	2.30	
Lock Indicator	5 VDC	.40	.60	1.00	mA**
Output Pin 8	9 VDC	3.00	4.30	6.90	
On (Logic "1")	12 VDC	6.10	8.50	13.00	
V_{OUT} Clamp to	15 VDC	10.40	14.00	21.00	
1.7 V	18 VDC	15.80	20.00	30.00*	
Save Indicator	5 VDC	.80	1.20	2.00	mA**
Output Pin 10	9 VDC	6.00	8.60	13.80	
On (Logic "1")	12 VDC	12.20	17.00	30.00*	
V_{OUT} Clamp to	15 VDC	20.80	28.00	30.00*	
1.7 V	18 VDC			30.00*	

*Indicates maximum allowable current drain of 30 mA.
Note: Limit output current to 30 mA max.

**Current drive balanced for equal brightness on red and green indicators.

in Fig. 10-3. When the car's ignition switch is turned on, the SENSE INPUT pin goes high and readies the SELECTED KEYS inputs (pin 3 through 6) to accept the unlocking sequence at input terminals I_1, I_2, I_3, and I_4. If the keys associated with those inputs are pressed in exactly in the right sequence, the LOCK CONTROL output goes high and lock relay RY1 is energized. At the same time, pin 8 goes

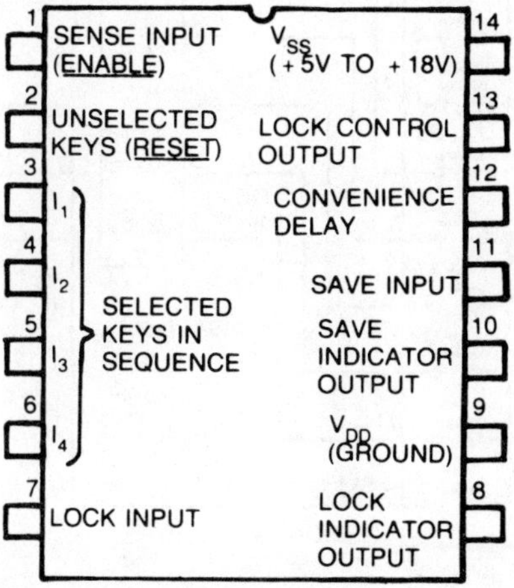

Fig. 10-1.

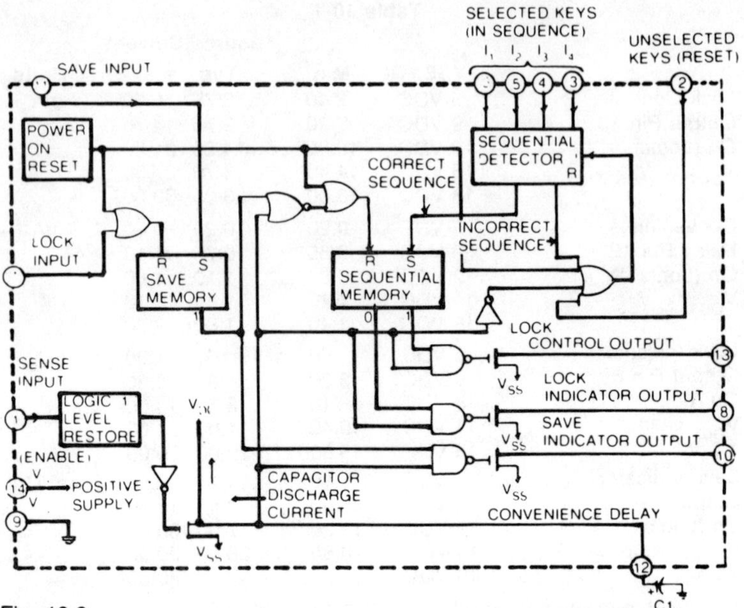

Fig. 10-2.

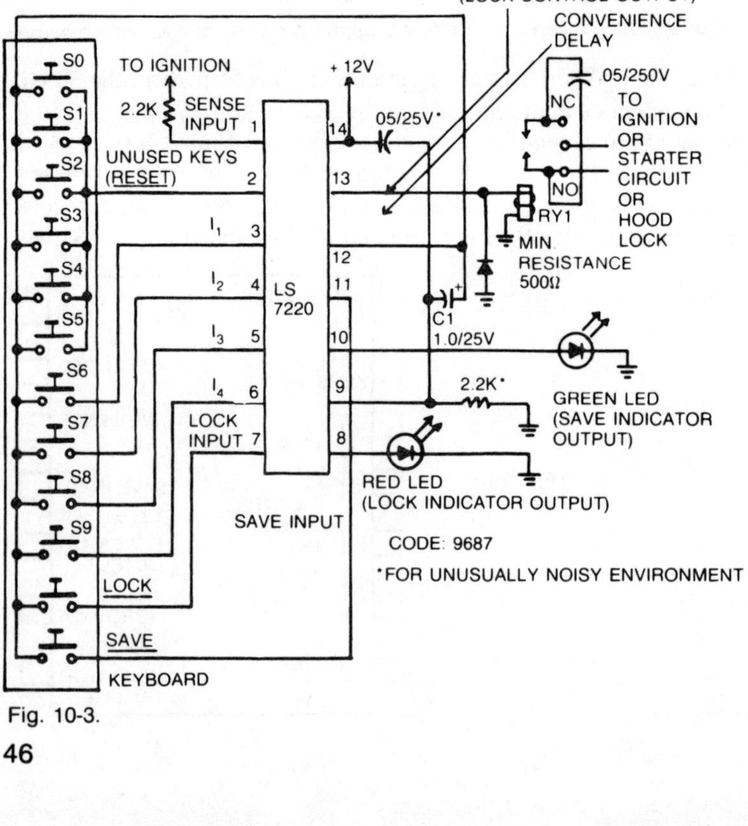

Fig. 10-3.

low, turning off the red LED to indicate that the circuit is unlocked. If the keys are pressed in any sequence other than that if I_1, I_2, I_3, and I_4, the internal sequential detector will reset and the entire keying sequence must be repeated.

At the moment that the sense input (ENABLE) pin goes high, the convenience-delay capacitor, C1, charges and delays any change in the states of the outputs when the state of the sense input changes from high to low. The delay is a function of the value of C1 and the supply voltage. When C1 is 1μF, the delay ranges from about 3 seconds with V_{SS} at 5 volts to 12 seconds when V_{SS} is 18 volts.

Normally, when the ignition is turned off, the sense-input pin goes low and the lock relay opens. The next time the ignition is turned on, the red LED LOCK indicator is turned on and remains on until the correct key code is punched in. The LOCK indicator goes out when lock relay RY1 is energized.

However, to hold or "save" the "unlocked" state when the ignition is turned off, even though the device is in the off or unlocked state, the driver first presses the SAVE key. That puts a logic "1" on pin 11 (SAVE INPUT) and sets the "save memory" flip-flop, preventing the sequential memory from resetting as the ignition is turned off. A logic "1" appears at the SAVE INDICATOR output pin (pin 10) and turns on the green LED. If the ignition is turned off while the green LED is on, all output states are preserved in the IC's memory and it won't be necessary to go through the input keying-sequence the next time the car is started. In effect, the "save" feature disables the security system so the car can be operated by valet parking and garage attendants. The security system can be switched out of the SAVE mode by pressing the LOCK key and then turning off the ignition for a time period longer than the convenience delay. That resets the save-memory flip-flop and arms the security system.

HIGH-SPEED ANALOG SWITCH

The Harris HI-201HS is a monolithic CMOS analog switch featuring very fast switching speeds and low on-resistance. It consists of four independently selectable SPST switches in a 16-pin DIP package. The device offers improved performance over earlier CMOS analog switches. Switching speed is 50 nanoseconds maximum and maximum on-resistance is 50 ohms. The wide analog-signal range of $\pm$ 15 volts makes the device ideal for sample-and-hold circuits, digital filters, op-amp gain-switching networks, and just about any other application where improved switching performance is required.

40-AMPERE PLASTIC SCR

A new line of TO-220 plastic-packaged SCR's (*S*ilicon-*C*ontrolled *R*ectifiers) with forward-current rates of 40 amperes—15 amps higher than previously available plastic devices—has been introduced by Motorola. The new devices have a surge-current rating of 400 amperes. That makes these SCR's particularly suitable for industrial applications.

The new MCR264 series includes four devices with voltage ratings between 200 volts and 800 volts as follows: MCR264-4 (200 PIV), MCR264-6 (400 PIV), MCR264-8 (600 PIV), and MCR264-10 (800 PIV).

Chapter 11

Compressors, Expanders, and Compandors

Since the early days of high-fidelity audio, engineers have worked to improve the realism and signal-to-noise ratio of both recorded and broadcast music. Recording engineers, however, often limit or compress the dynamic range, and broadcasters limit or compress the signal amplitude, of that music. That is done to prevent overloading a tape or overcutting a record, and to prevent overmodulation. However, those same efforts cause the full dynamic range of the original music to be lost to you—if your playback system does not include a *dynamic volume expander.*

THE SIGNETICS NE570 COMPANDOR

You can build a professional-quality expander, compressor, or compandor (a combination compressor and expander circuit) for your hi-fi system by using circuits designed around the Signetics NE570 IC compandor. The pin-out of the IC is shown in Fig. 11-1. As a compressor, the device provides a 2:1 compression ratio—for example, a 100-dB dynamic range of +20 dB to −80 dB is compressed into a 50-dB (+10 to −40 dB) range as shown in Fig. 11-2. As an expander, it has a 1:2 expansion ratio, taking the +10 to −40 dB compressed signal and restoring its original full dynamic range.

A compandor can be used for noise reduction. In that application, the signal is compressed before noise can be introduced,

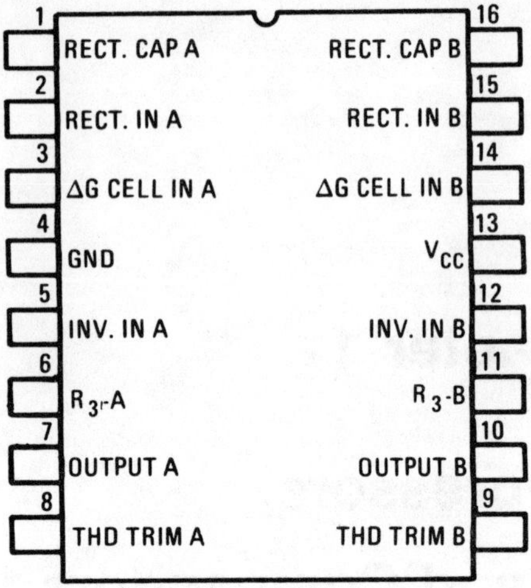

Fig. 11-1.

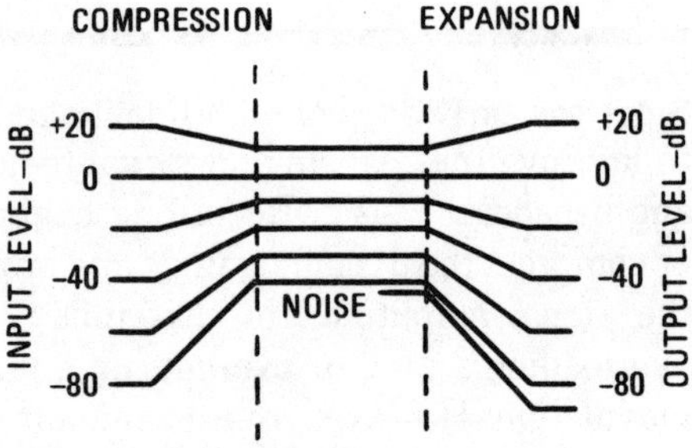

Fig. 11-2.

and expanded afterwards. Figure 11-2 shows how that method of companding (compressing and then expanding) can improve the signal-to-noise ratio by about 45 dB.

A block diagram of one half of the NE570 is shown in Fig. 11-3. Each half of the IC consists of a full-wave rectifier, a variable-gain cell (ΔG), an op amp, and a biasing system. The full-wave rectifier and an external capacitor (tied to the RECT CAP terminal) detect the average value of the input signal. The rectifier output current (I_G) controls the gain of the variable-gain cell. Therefore, the gain of that section of the circuit is proportional to the average value of the input-signal voltage. The ΔG output current, I_{OUT}, is fed to the inverting input of an on-chip op amp that is biased at V_{REF}. That reference voltage is 1.8 volts and is provided by a very stable internal low-noise source. (That internal precision voltage-source also biases the THD TRIM circuit used for temperature compensation.)

The speed with which the circuit gain can follow changes in the amplitude of the input signal depends on the value of the external capacitor (the one attached to the RECT CAP terminal). A small capacitor will provide fast attack and fast decay times, but may not provide enough low-frequency filtering. In that case, residual low-frequency signal components will appear on I_G and will modulate the signal passing through the variable-gain stage. That results in third-harmonic distortion, so there must be a compromise between fast response and distortion.

The expander's output is determined by the DC gain provided by the op amp. The output is related to the internal reference voltage and also to biasing resistors R_3 and R_4 as expressed by the equation: $V_{DC\,OUT} = (1 + R3/R4)\,V_{REF}$. When V_{CC} (the supply voltage) is higher than 6 volts, R_4 should be shunted with an external resistor to bias the output up to ½ V_{CC}.

Resistor R_3 is brought out from the op amp summing node and is used when you want expander or compressor gain to be set solely by on-chip components. You can adjust that gain to your needs by placing external resistors in series with R_3. You can also connect an external resistor across R_4 to change the bias to any value desired.

The Basic Expander

Figure 11-4 shows the circuit of a basic expander. Input signal V_{IN} is applied to the rectifier and ΔG stage inputs in parallel. The expander can handle a signal input up to 3 volts peak. Rectifier input current can be as high as 300 μA, while the input to the ΔG

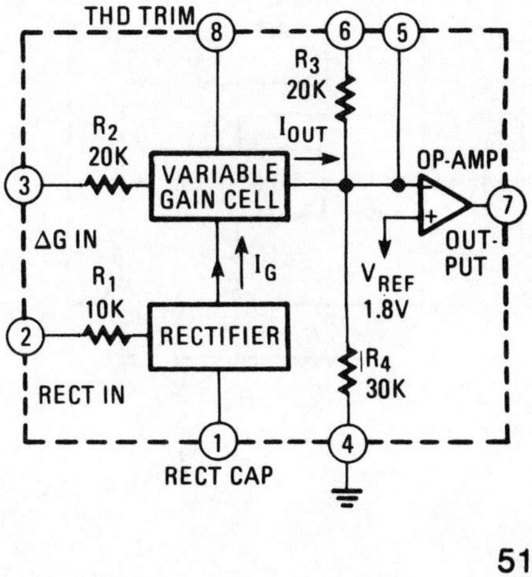

Fig. 11-3.

stage should be limited to 140 μA. If the compandor will see input signals greater than +2.8-volts peak, use suitable resistors in series with R_1 and R_2 to limit currents to the specified values.

Voltage offsets in the ΔG stage can cause distortion: primarily even harmonics. The THD TRIM pin permits a compensating external voltage to be applied to neutralize the effect of the offset voltages. A voltage divider composed of a 20K pot and series resistor R is connected between V_{CC} and ground as shown in Fig. 11-4. A 6.2K resistor is connected to the THD TRIM pin. The value of resistor R is selected to develop 3.6 volts at the high end of the pot.

In Fig. 11-4 coupling capacitors are shown in series with both the rectifier and ΔG stage inputs. However, R_1 and R_2 can be tied together and connected to the signal input through a single cou-

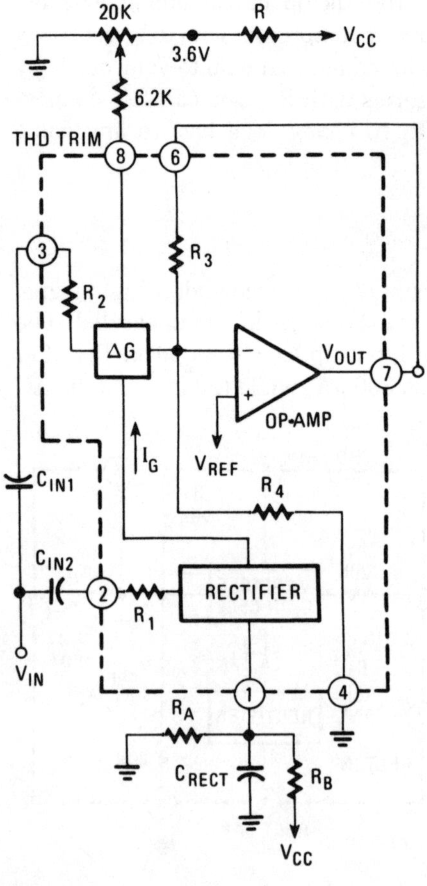

Fig. 11-4.

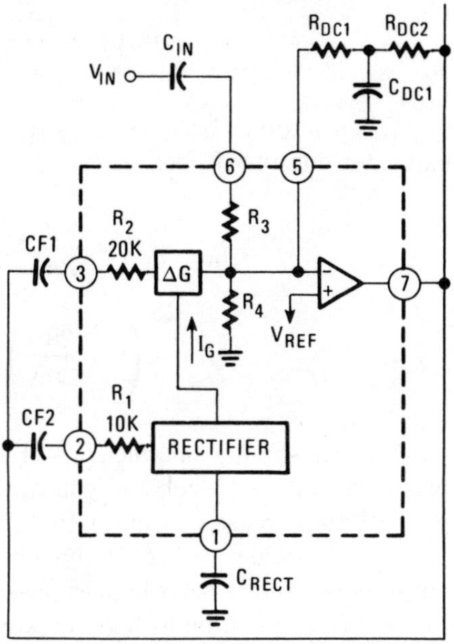

Fig. 11-5.

pling capacitor. In that case, though, tracking at low input-signal levels will be degraded.

The comparator transfer-tracking tends to be a linear 2:1 ratio down to a very low input-signal level. Then, tracking may deviate in either direction from the normal 2:1. Either resistor R_A or R_B (but not both) may be needed to adjust transfer linearity. To correct low-level tracking error, select a suitable value for R_A ranging from around 1 megohm to 100K or for R_B between 250K and 5 megohms.

The Basic Compressor

Figure 11-5 shows that the dynamic compressor is essentially an expander inserted in the feedback loop of an op amp. The inputs of the ΔG stage and the rectifier are tied to the op amp output. The variable-gain stage is set to provide AC feedback only; DC feedback is provided by an external low-pass network composed of R_{DC1}, R_{DC2}, and C_{DC1}. The sum of the values of the two feedback resistors determines the bias at the op amps output. The output voltage $V_{DC\ OUT}$ can be written:

$$V_{DC\ OUT} = 1 + \frac{R_{DC1} + R_{DC2}}{R_4} V_{REF}$$

$$= \left(1 + \frac{R_{DC\ TOT}}{30K}\right) 1.8V$$

When internal bias resistors R_3 and R_4 are used alone, the expander output will be:

$$V_{DC\ OUT} = 1 + \frac{R3}{R4} V_{REF}$$

$$= \left(1 + \frac{20K}{30K}\right) 1.8V = 3.0V$$

You can shunt a suitable resistor across R_4 to raise the output bias to the desired level; and you can connect a resistor in series with R3 to increase op amp output gain.

For the widest possible dynamic range, the compressor's output-level should be as high as possible. Therefore, the input to the rectifier should be as high as possible without exceeding the +300 μA peak-current limit. If the average input-signal level is low, a higher output can be obtained by using a shunt resistor to reduce the effective value of R_3 or by using an external series resistor to increase the effective value of R_2. Note well that a reduction in the effective value of R_3 reduces the circuit's input impedance.

A High-Fidelity Compressor

Figure 11-6 shows a circuit for a hi-fi dynamic compressor that would make an ideal accessory for your tape-recording setup. It features high gain and wide bandwidth. Its external rectifier-capacitor (C9) is not grounded. Instead, it is connected to the output of an op amp network (IC1-a and IC1-b) to shorten the compressor attack-time at low signal-levels. (The attack time of the basic circuits in Figs. 11-4 and 11-5 is relatively long.) That external op amp is used to provide improved high-frequency gain.

Diode D3 and D4 clamp the compressor output to a 7-volt peak-to-peak swing. That is necessary at times when the compressor is operating near maximum gain—as with a small signal input—and is suddenly hit with a high-level signal. Normally, the output would swing from V_{CC} to ground and would overload the circuit the compressor was feeding—a tape recorder for example. The attack time and the time it takes for the compressor to recover from an

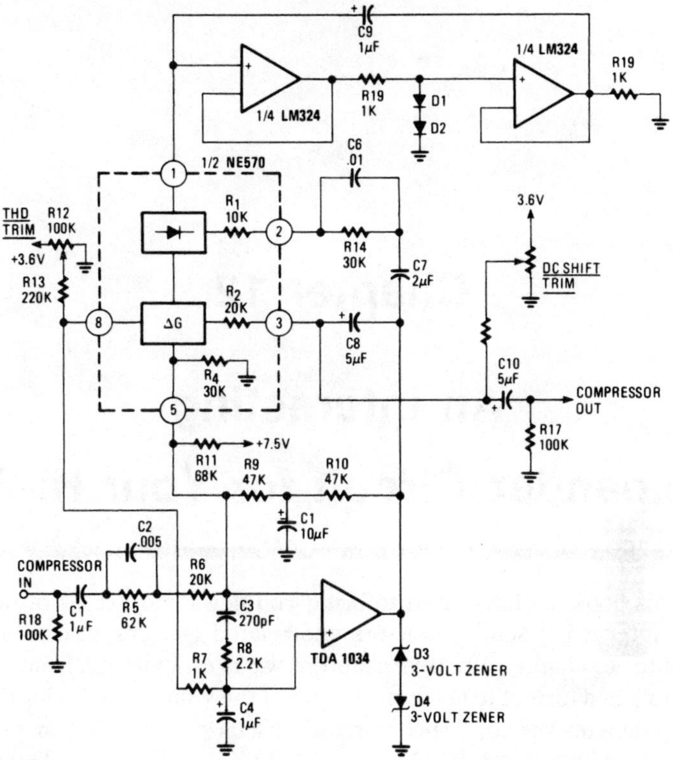

Fig. 11-6.

overload depend on the value of C9. A value of about 1 μF is a good compromise.

Breathing

Even some of the best broadcast-quality compressors have been said to have a problem with *breathing*. That term refers to slow cyclic variations in background level that can be heard as the compressor changes gain. Breathing is minimized in this circuit by high-frequency pre-emphasis networks C2-R5 and C8-R14. Naturally, the expander should have a de-emphasis network to complement the compressor's pre-emphasis network. We'll take a look at the expander circuit, before we go on to other things, next month.

This material was abstracted from the Signetics *Compandor Product Guide* from the Analog Division, **Signetics Corp.**, PO Box 409, Sunnyvale, CA 94086.

Chapter 12

An Interesting Expander Circuit for Your Hi-Fi

In this book we have tried to bring you news and details of new and interesting semiconductors and related matters without any chapter-to-chapter. In this chapter, we are deviating from the established format to show the circuit of the volume expander that complements the volume compressor discussed in the last chapter.

The hi-fi expander in Fig. 12-1 includes de-emphasis and attack and decay times to complement the compressor. Again, an external op amp is substituted for the one in the Signetics NE570 compandor IC. In this case, the circuit performance demands a slew rate better than the 0.6 volts-per-microsecond provided by the NE570 op amp. The expander, like the compressor, has a unity-gain level of 0 dBm. Adjustments are provided for harmonic distortion and DC shift.

Make the THD TRIM adjustments first using a 10-kHz 0-dBm (774.6 mV) signal. Adjust the DC SHIFT TRIM for minimum envelope bounce when tone bursts are fed through the circuit.

Performance is reported to be spectacular when the expander is applied to consumer tape recorders.

OP AMP AND VOLTAGE-REFERENCE IC

The National Semiconductor LM10-series comprises a group of monolithic linear IC's consisting of a precision voltage reference, an adjustable reference buffer, and an independent high-quality op-amp. The device has an absolute maximum supply-voltage rating

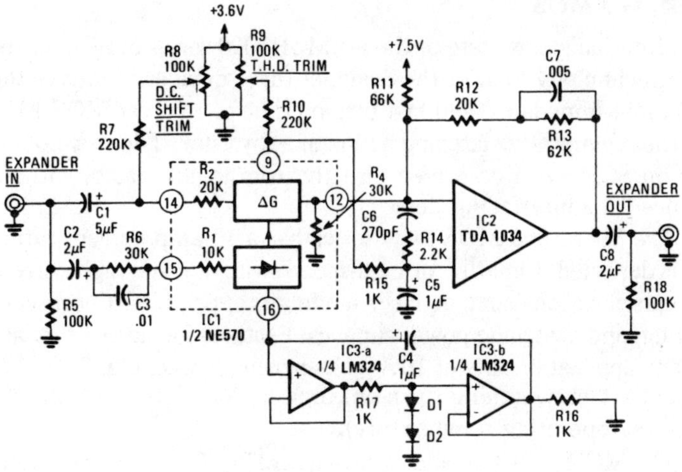

Fig. 12-1.

of 45 volts for the LM10, LM10B, and LM10C; and 7 volts for the LM10BL and LM10CL. Supply voltage can be as low as 1.1 volts. Typical supply current is 270 µA.

The complementary output stage in the independent op amp is capable of swinging to within 15mV of the supply voltage or delivering + 20-mA output current with + 0.5 volt saturation. The reference output can be as low as 200 mV with 0.1% regulation.

The LM10's ability to operate from a single power supply with high output-current capabilities make it a very versatile general-purpose device. It can operate in a floating mode, independent of fixed supplies so it can be used in such applications as remote amplifiers for vibration sensors, flame detectors, resistance thermometers, optical pyrometers, and logarithmic light sensors. Its ability to operate from a wide range of voltages and currents makes the LM10 applicable in a wide range of voltage and current regulators.

The LM10 features such characteristics as 2.0 mV (max) input offset voltage, 0.7 nA (max) input offset current, 20 nA maximum input bias current, and offset-voltage drift of 2 microvolts-per- °C. The device comes in a metal can and an 8-pin dual-in-line package.

Full electrical specifications, including 33 typical-performance-characteristics charts are in the 15-page data booklet along with 33 schematics illustrating typical applications.—**National Semiconductor**, Literature Dept., M/S 16251, 1090 Kifer Rd., Sunnyvale, CA 94086.

SCRS GO MOS

Motorola's new state-of-the-art MCR1000 series of SCR's uses MOS technology to offer the designer three devices that have the high input-impedance and fast turn-on time of a power MOS-FET and the regenerative latching action of a thyristor. The new 200-ns turn-on SCR's can be driven directly from logic circuits without the need for interfacing devices.

The MCR1000 MOS SCR series has a 15-amp current rating and is designed primarily for high-speed switching and high-current pulse applications such as laser modulators, printers, fluorescent lighting, and switching power supplies. Features include: A forward voltage application rate of 1000 volts-per-microsecond at T_j = 125 °C and a fast switching (turn-on) time of 200 ns at T_j = 25 °C. (T_j is the operating temperature).

The MCR1000-4, -6 and -8 devices are in TO-220 packages and feature peak forward blocking voltages of 200, 400, and 600 volts, respectively.

1-WATT POWER MOSFETS

International Rectifier has introduced a family of four new n-channel transistors using the company's HEXFET technology. A particular advantage of the devices is their low on-state resistance—about half that of many similar devices. Designated IRFD1Z0 through IRFD1Z3, the four devices have typical rise and fall times of 15 ns and 10 ns, respectively when I_D (continuous drain current) is 0.25 A. On-state resistance is 2.4 ohms for the IRFD1Z0 and -IZ1 and 3.2 ohms for the IRFD1Z2 and -1Z3.

PHASE-LOCKED-LOOP

The Exar XR-2213 is a highly stable phase-locked-loop IC designed for use in tone-detection applications and in FM detectors, tracking filters, and frequency synthesizers. Its circuit consists of a high-stability VCO, an input preamp, loop phase detector, lock phase detector, and a high-gain voltage comparator.

The XR-2213 has an operating frequency range of 0.02 Hz to 300 kHz and a preamp input voltage range of 2 mV to 3 V rms. The VCO and comparator supply TTL-compatible output signals and both Q and $\overline{Q}$ outputs. The 5-page data sheet includes design equations and examples along with circuit diagrams that illustrate various applications. Sample circuits include a tone detector, and a frequency synthesizer developing a 40-kHz output from a precision 10-kHz input signal.—**Exar Integrated Systems**, 750 Palomar Ave., Sunnyvale, CA 94088.

SEMICONDUCTOR SUBSTITUTION GUIDE

The Power Semiconductor Cross-Reference Guide, from Westinghouse, covers all rectifiers, transistors, SCRs and assemblies available from the company. It includes more than 10,000 JEDEC, Westinghouse, and competitive part numbers from 23 manufacturers. The guide is free.—**Westinghouse Corp.**, Semiconductor Div., Youngwood, PA 15697.

LINEAR-IC GUIDE

Audio/Radio Products is a 19-page booklet prepared to introduce National Semiconductor's line of linear ICs for use in radio and audio products. Section 1 contains a 69-entry glossary of terms, definitions of maximum ratings and other IC parameters; and a check-list of do's and don't's for optimum performance from ICs. This check-list covers everything from power-supply selection and bypassing to circuit layout and heatsinking. Section 2 is the selection guide, listing devices under such application headings as tape processors, audio drive ICs, stereo decoders, audio power-amplifiers, and tone-control amplifiers. Key technical specifications are listed in tables and charts. Section 3 is a pocket containing about 2 dozen data sheets of typical devices that are covered in the previous section.

The glossary and check-list alone make this booklet a valuable addition to the libraries or technical data files of engineers, designers and others interested in IC's for radio and low-power audio applications.—**National Semiconductor Corp.,** 2900 Semiconductor Drive, Santa Clara, CA 95051.

Chapter 13

An Automatic Brightness Controller

A unique device that is quite versatile in applications that require the automatic adjustment of the brightness of a display in relation to variations in ambient light, is the new Optron OPL 100 automatic brightness controller. That device from TRW Optron can operate from a supply voltage of from 4.5 to 24 volts DC, and can source 50 mA or sink 20 mA to control the brightness of various types of displays from 0 to 100 percent. The display types that it can control include vacuum fluorescent, incandescent, and LEDs. The OPL 100 can be used to control TV screens, instrument displays in aircraft and automobiles, industrial light levels, and outdoor lighting for signs and security purposes. The device can also be used to keep the indoor light level constant, regardless of variations in light coming in from outdoors.

Pulse-width modulation is the key to the controller's operation. The duty cycle of the output is proportional to the ambient light level and is the basis for controlling external displays. Ambient light levels are detected by a 2500 square mil (0.0025 square inch) photodiode that is built into the device.

Figure 13-1 shows a block diagram of the OPL 100, an 8-pin IC. Figure 13-2 shows how it's used to control the brightness of LEDs, 7-segment numeric displays, and vacuum-fluorescent displays. In each case, the displays are driven by the pulse-width-modulated output from pin 7. Potentiometer R1 is the external adjustment for light sensitivity level. The resistance needed

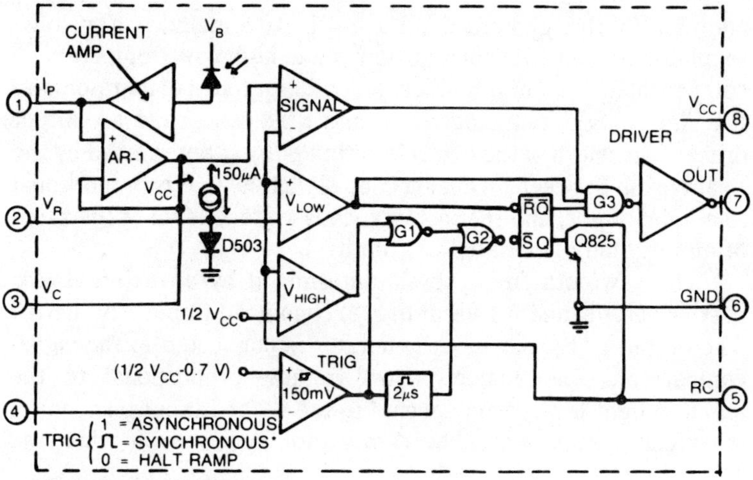

*TRIGGERS ON LEADING EDGE

Fig. 13-1.

depends on the ambient light level, and on the display brightness desired—it ranges between 25K and 200K.

In Fig. 13-2(a), the device controls the brightness of a standard LED. With a 12-volt DC supply, the controller furnishes a maximum of 32 mA to the LED. Figure 13-2(b) shows how the controller is connected to a numeric display through a BCD-to-7-segment latch/decoder/driver like the CD4511. Several drivers can be used and multiplexed by connecting the strobe or enable line of each driver to pin 4 of the controller through suitable logic circuitry.

In Fig. 13-2(c), we see how to connect the OPL 100 to a vacuum-fluorescent display device to control display brightness between 0% and 100% intensity. Resistor R2 (shown as dashed lines) is needed in this application to prevent the display from going completely off when the sensor is in total darkness. The OPL 100 should be mounted so that its photodiode sensor receives light from the direction from which the display will be viewed. Potentiometer R1 should be adjusted for a satisfactory display brightness in normal ambient light. Now, the controller will detect changes in ambient light and adjust the brightness of the display. Because of that, the display brightness will appear to remain constant.

HOW THE OPL 100 WORKS

Referring to the block diagram in Fig. 13-1, we see that, in addition to the photodiode, the OPL 100 includes a voltage regulator, a constant-current source, a high-gain temperature-compensated current amplifier, an op amp, four comparators, an R-S latch, logic including NAND and NOR gates, and an output driver. The width of the device's output pulse is determined by the relationship between an analog control-voltage that is proportional to ambient light, and the instantaneous voltage level of the ramp of an internally generated sawtooth.

The sawtooth frequency is determined by an external R-C network (100K and 0.1 μF in Fig. 2) connected to pin 5 of the IC. The output of the sawtooth generator is connected to the signal comparator. The analog control-voltage—proportional to the ambient light level—is connected to the second input terminal of the signal comparator. The comparator switches on when the

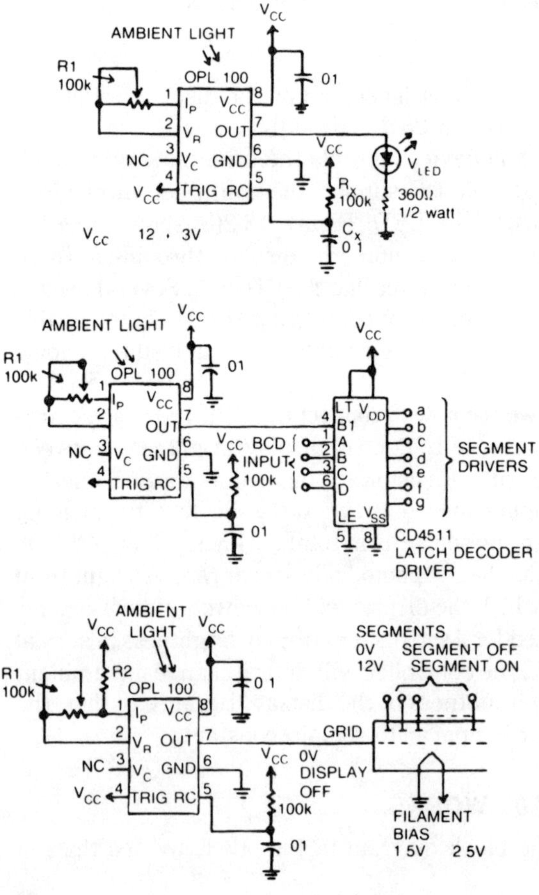

Fig. 13-2.

control voltage is higher than the instantaneous sawtooth-ramp voltage, and switches off when the control voltage is lower.

The "on" portion of the output pulse and the sawtooth ramp start at the same time. The output-pulse width continues to rise until the ramp voltage is higher than the control voltage V_C. At that point, the signal comparator switches off. Since the control voltage rises with ambient light, we see that as light level increases, V_C rises, causing an increase in the width of the output pulse. That causes an increase in the display brightness.

The sawtooth frequency is approximately $1.44/(R_xC_x)$ where R_x and C_x are the values of the external timing components. The maximum recommended value for R_x is 100K. To obtain an asynchronous (free-running) sawtooth, connect pin 4 of the IC to V_{CC} as shown in the applications in Fig. 13-2. That lets the ramp voltage to rise to ½ V_{CC} and then fall to 0.7 volt.

To obtain a synchronous (externally triggered) sawtooth voltage, a sync (trigger) pulse must be applied to pin 4 of the IC. The ramp is synchronized with, and triggered on, the positive-going edge of the sync pulse. The sync-pulse amplitude must be greater than $½V_{CC} - 0.7$ volt. The free-running frequency of the internal ramp generator—as set by the R-C network tied to pin 5—should be slightly lower than the frequency of the external-sync signal.

Let's now look at each pin of the OPL 100 and its function.

Pin Functions

Pin 1 is the source of amplified photocurrent through an open-collector PNP current source. The variable resistor between pins 1 and 2 is the load resistor for the current source. Photocurrent I_P is ½ mA under normal room light. In addition to providing the dimming function, pin 1 can be used for on-off control of the display. Connecting pin 1 to V_{CC} forces the output terminal (pin 7) to a logic-1 (V_{CC} − 2 volts), driving the displays to full brilliance. Connecting pin 1 to ground forces pin 7 to logic-0, and turns the display off.

Pin 2 is the source of reference voltage V_R (approximately 680 mV) set by a constant current of approximately 150 μA through diode D503. That voltage determines the lower voltage extreme for the sawtooth waveform. Pin 2 is the tie-point for the lower end of the sensitivity control.

Pin 3 is the source of a buffered analog control-voltage that is proportional to the incident light level. In applications where several OPL 100s are used in a system, control voltage V_C from the main unit is applied to pin 1 on the remote OPL 100s so their duty cycles will track the duty cycle of the main unit.

Pin 4 is the trigger pin. Its external connections determine the operation of the sawtooth generator. Connecting it to $\frac{1}{2}V_{CC}$ - 0.7 volt causes the sawtooth generator to free-run in an synchronous mode with the upper voltage-extreme being approximately $\frac{1}{2}V_{CC}$ (reference for the V_{HIGH} comparator) and the lower extreme approximately 680 mV (reference for the V_{LOW} comparator).

When synchronous operation is needed—as for multiplexing digits—a positive-going pulse is applied to pin 4. The leading (rising) edge should coincide with the beginning of each digit's "enable" time. That causes the sawtooth to stop rising, discharge to its lowest level (V_R), and then begin its ramp-up.

Grounding pin 4 stops the sawtooth and causes pin 5 to rise to V_{CC} and remain at that level until the ground is lifted and a logic-1 or a pulse is applied (for asynchronous or synchronous operation, respectively.)

Pin 5 is the tie-point for R_x and C_x, which control the frequency or repetition rate of the sawtooth. The average discharge-time is 4 μs for a 0.1 μF timing capacitor. The sawtooth repetition rate is determined from the formula $1.44/(R_xC_x)$, where C_x is in μF and R_x is in megohms.

Pin 6 is the ground terminal for the device.

Pin 7 is the output of the driver amplifier. It will directly drive the grid of a vacuum-fluorescent display. In addition, it can source 50 mA at a minimum voltage of V_{CC} – 4 volts, and can sink 20 mA with a maximum of 0.4 volt. Incandescent lamps and 7-segment numeric displays require additional interfacing components.

Pin 8 is the input terminal for supply voltage V_{CC}, which can range from 4.5 to 24 volts DC. Supply current is around 12 mA and is relatively constant over the operating-voltage range when in total darkness. Supply current increases with light intensity.

AMBIENT-LIGHT CONTROLLER

Figure 13-3 is a circuit you can use to keep the ambient indoor light level constant. For example, it can be used when you must increase the artificial indoor lighting level to compensate for a decrease in incoming natural light. The brightness of an incandescent light source is controlled by varying its duty cycle through a power triac. The triac, in turn is controlled by the OPL 3010 optoelectronic

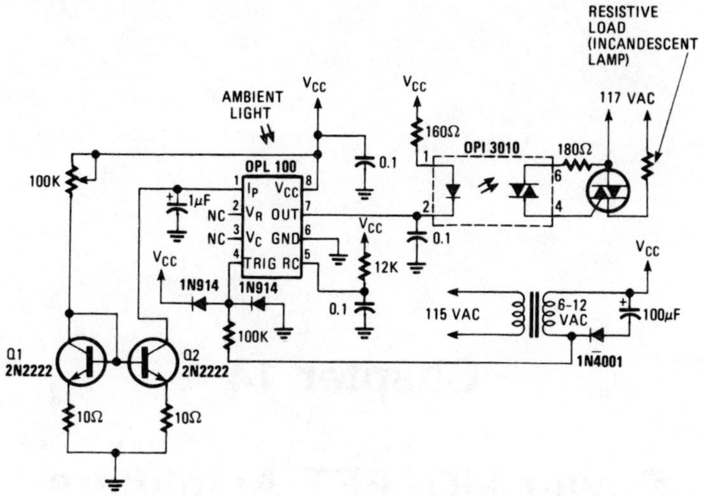

Fig. 13-3.

triac-driver. When that is done, the OPL 100 controller must be synchronized with the AC power line to ensure proper triggering of the power triac. That is accomplished by developing a sync pulse from the secondary winding of a small 6—12-volt transformer and a 1N4001 diode in the circuit between pin 4 and V_{CC}.

The 100K pot sets the ambient light level by establishing a constant current through transistor Q1. The photodiode sensor in the OPL 100 then sets the output duty-cycle to a point that produces enough light to generate an amplified light-current (I_P) at pin 1 that is equal to the current through Q1. As the incoming light fades, the output duty cycle is altered to increase the "on" time of the indoor incandescent lighting to keep the ambient light level constant.

This item was prepared from material supplied by TRW Optron. If you'd like additional information on the OPL 100 Automatic Brightness Controller, write to TRW Option, 1201 Tappan Circle, Carrollton, TX 75006 and ask for Product Bulletin 2091 and Application Bulletin 115.

Chapter 14

Power MOSFET Amplifiers

Some of the newer hi-fi amplifiers are designed around power FETs in the output stage. That's because they offer low harmonic distortion, extended SOA (*S*afe *O*peration *A*rea), and infinite current gain. The design, construction, and performance of such an amplifier is covered in a recent Intersil applications note. Its schematic diagram is shown in Fig 14-1. The amplifier delivers 50 watts into an 8-ohm load. Its frequency response is ±3-dB from 20 Hz to 100 kHz while distortion is less than 0.25% from 20 Hz to 20 kHz.

The amplifier's design is unconventional in several respects: First, there are no rolloff or compensation capacitors. Instead, rolloff is provided by the input capacitance of the power MOSFET's (Q8 and Q9) that are arranged in a quasi-complementary output stage. The driver stage uses bipolar transistors and provides the usual voltage amplification and phase-splitting functions. Finally, the positive power-supply line is boot-strapped to insure that the positive and negative half-cycles of the output signal are equal and maximized.

The input stage is designed around Q3, a U402 dual n-channelJFET in a differential configuration. A pair of J113 JFET's (Q1 and Q2) provide a constant current for the U402. That current controls the bias current of the output stage and is set by R18, a 10-turn 1K trimmer potentiometer. Bipolar transistor pairs Q4-Q6 and Q5-Q7 turn on in proportion to the input signal and provide

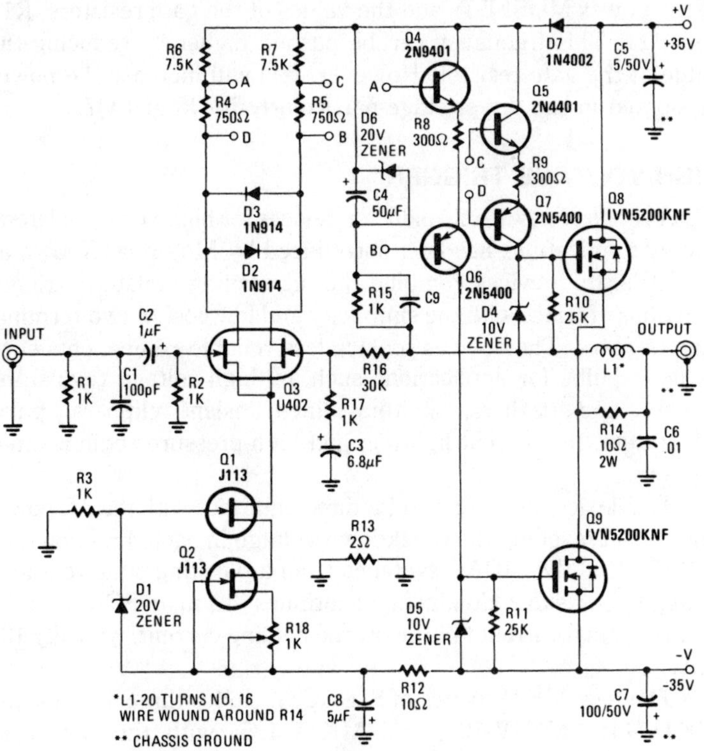

Fig. 14-1.

drive current for output transistors Q9 and Q8, respectively. The drive current flows through R10 and R11 and develops the necessary gate-source voltage for the output transistors. The power gain of the output stage is quite high because of the comparatively high value of the resistors connected between gate and source.

The boot-strapped positive power-supply line, mentioned earlier, is needed because if Q8 is to clip at the level of the positive power-supply line, its gate voltage must be driven beyond that point. Thus, a voltage higher than ⅙ must be supplied to the input and driver stages. Instead of using an external power supply with an output of more than 35 volts, the designers elected to bootstrap the ⅙-volt supply; that bootstrap circuit consists of D7, C4, D6, C9, and R15.

The closed-loop gain is determined by resistors R16 and R17. Capacitor C3 compensates for any offset in the output. In this design, the open-loop gain is set at 20 kHz by the gate capacitance of the power MOSFETs and the values of the gate resistors, R10 and R11. This frequency can be pushed higher by reducing the value of the gate resistor. However, this will increase the power dissipated in the driver stage-particularly in Q6 and Q7.

HIGH-VOLTAGE TRIGGER

A series of new semiconductor devices for high-voltage bilateral trigger applications has been introduced by Motorola. Known as SIDAC's, the devices combine the high-voltage bilateral trigger functions of triacs with the simplicity and low cost of two-terminal diac triggers. The devices replace less reliable components such as neon bulbs for applications such as: high-voltage regulators, strobes and flashers, ignitors, line-transient clippers, pulse generators, fluorescent lighting, and high-pressure sodium-vapor lighting.

The devices are designed for direct interface with the AC power line. Upon reaching the breakdown voltage in each direction (104 to 135 volts) the SIDAC switches from a blocking state to a low-voltage "on" state. Conduction continues—as in an SCR-until the main terminal current falls below the holding current, typically 100 mA.

The new MK1V series of SIDACs includes three devices, the MK1V-115, MK1V-125, and MK1V-135; with the last three numbers indicating the maximum repetitive break-over voltage. Figure 14-2 shows two SIDAC circuits; Fig. 14-2(a) shows how electronic equipment can be protected against line-voltage transients and Fig. 14-2(b) shows a SIDAC as a two-terminal trigger for a strobe or flasher application.

The minimum breakover (breakdown) voltages are 104, 110, and 120, respectively for the three devices. Off-state repetitive voltage is ± 90 volts and repetitive peak off-state current is 10 μA. Instantaneous on current is 1.0 amp RMS for each device. Maximum breakdown current (60-Hz sinewave) is 200 μA. Forward "on" voltage is typically 1.1 while dynamic holding current is 100 mA.

TRANSISTOR DATA BOOK

Field-Effect Transistors-Selector Guide and Cross Reference is a 32-page booklet chock full of valuable information on FETs. It opens with a brief description of all types of field-effect transistors and then comparing FETs to bipolar devices and then JFETs versus

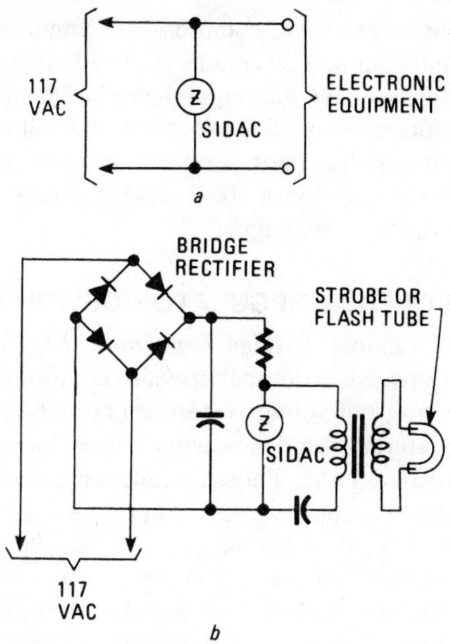

Fig. 14-2.

MOSFETs. This is followed by many listings of pertinent characteristics of JFETs, single- and dual-gate MOSFETs, dual-gate DMOS as well as power MOSFETs.

A numerical index lists approximately 70 devices whose characteristics are listed in tables depending on their classification as Switches and Choppers, Power MOS, Low-Frequency Low-Noise Amplifiers, or High-Frequency Amplifiers. A Cross-Reference and Interchangeability Guide matches industry type numbers of approximately 1900 FETs with equivalent types from Motorola. The book concludes with illustrated FET applications that include an ionization-chamber smoke detector and a 500-MHz dual-gate MOSFET amplifier that is suitable for service in the front-end of a communications receiver.—**Motorola Semiconductor Products,** PO Box 20912, Phoenix, AZ 35036.

HIGH-POWER 900-MHz TRANSISTORS

Motorola has introduced three new 24-volt DC, 900-MHz power transistors. The line includes the MRF890, a 2-watt, 9.0-dB minimum-gain predriver; the MRF892 14-watt, 8.5-dB driver; and the MRF894 30-watt, 7.0-dB final amplifier. A chain of those three devices is capable of boosting a 100-mW UHF input signal (804 to 960 MHz) to 30 watts output. The new transistors are intended for large-signal, common-base amplifier applications in industrial and commercial cellular-FM radiotelephone equipment. All three devices have guaranteed gain performance at 900 MHz, collector efficiencies of 55% minimum, and will withstand a 30:1 VSWR load mismatch at their rated voltages and power outputs. The MRF390 is packaged in a 305-1 case: the other two devices in the family in 319-04 packages.

VOLTAGE REGULATOR DESIGN DATA

Linear Voltage Regulators, (Application Note APN-27) is a 16-page booklet that provides detailed design information on a wide variety of voltage-reference circuits and voltage regulators. The material includes schematics, performance data, and descriptions of 18 circuits. Those circuits include band-gap voltage references, Zener-diode references, series and shunt regulators, and positive and negative voltage regulators. The designs include information on such parameters as temperature stability, noise voltages, and line and load regulation.—**Interdesign, Inc.,** 1255 Reamwood Ave., Sunnyvale, CA 94086.

Chapter 15

Two Versatile Voltage References

Voltage references are used to provide a constant output-voltage regardless of changes in input voltage, load current, or temperature. The REF-01 and REF-02 are monolithic voltage-reference IC's from PMI (Precision Monolithics Inc., 1500 Space Park Drive, Santa Clara, CA 95050). The REF-01 develops + 10 volts and the REF-02 + 5 volts. Both devices feature excellent temperature stability, low noise, low power consumption, and high load-driving capability. They are available in three package-types: TO-99, 8-pin hermetic DIP, and 8-pin epoxy mini-DIP. Applications include voltage references in analog/digital converters and digital/analog converters, precision current-sources, use in panel meters, cold-junction thermistor compensation, and calibration standards.

The excellent temperature-stability of the output voltage of these voltage references is due to the application of two theorems in band-gap semiconductor physics. One theorem states that the base-emitter voltage (V_{BE}) of a transistor is dependent on the collector-current and has a negative temperature coefficient. The second theorem states that the difference in the V_{BE} of two transistors operating with different currents has a positive temperature coefficient.

When the difference in the two V_{BE}s (ΔV_{BE}) is amplified and added to the V_{BE} of one of the transistors, a voltage reference with a zero temperature-coefficient results if the sum of the two terms equals 1.23 volts—the band-gap voltage of a silicon junction at 0 °K or −273 °C. That 1.23 volts is amplified to produce stable outputs of + 5.000 volts in the REF-02 and + 10.000 volts in the REF-01.

In addition to the stable + 5-volt reference output, the REF-02 provides an additional output voltage that is linearly proportional to the ambient temperature. That voltage, V_{TEMP}, is taken from the junction of R1 and R2. Figure 15-1 is a simplified schematic of the REF-02. The REF-01 is similar but it does not bring V_{TEMP} out to a terminal pin.

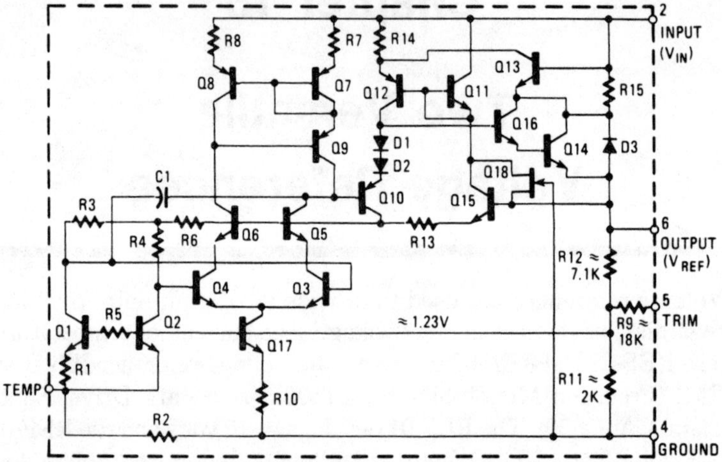

Fig. 15-1.

The REF-02 generates a ΔV_{BE} of 72 mV at 25 °C. That's done by making the current density (current/area) through Q2 sixteen times greater than that through Q1; while transistor Q2 has four times the emitter current of Q1, Q1 has four times as much emitter area as Q2. A voltage equal to ΔV_{BE} (72 mV) is developed across R1 and is amplified 8.75 times and then added to the V_{BE} of Q2 to become V_Z—1.23 volts with a zero temperature-coefficient. As a proof: V_{BE} of a silicon transistor is 0.6 volts (600 mV) at 20 °C and 8.75 × 73 mV = 630 mV, and 600 + 630 = 1230 mV or 1.23 volts.

The −2.1 mV/°C temperature coefficient (TCV_{BE}) is canceled by the + 2.1 mV/°C (TCV_{TEMP}) to produce a V_Z of 1.23 volts. That value is amplified by 4.06 to produce a V_{REF} of 5 volts in the REF-02, or amplified by 8.13 to produce a V_{REF} of 10 volts in the REF-01.

In the next column we'll look at how we can use either voltage-reference IC, along with a handful of other parts, to build an electronic thermometer. Direct temperature readings (in either °F or °C) can be displayed on a voltmeter (digital if desired).

PROGRAMMABLE QUAD COMPARATOR

The new National Semiconductor LP165/LP365 quad comparators feature low-power operation (10 μW per comparator) over a wide range of supply voltages. The four comparators on each chip can be programmed simultaneously for various supply currents, input currents, response times, and output-current drives. The family operates from a single supply with outputs ranging from 4 to 36 volts, or from split-power supplies delivering ± 2 to ± 18 volts DC. The comparator outputs are compatible with DTL, TTL, MOS, CMOS, and most other logic families.

Applications include battery power circuits, threshold detectors, zero-crossing detectors, VCO's, multivibrators, and A/D and V/F converters.

HIGH-TEMPERATURE RECTIFIERS

Four new Schottky rectifiers that have a 175 °C junction capability have been added to Motorola's product line. The higher-temperature 65- and 85-ampere, 35- and 45-volt diodes offer a greater temperature-margin in switching-power-supply design than previous 150 °C devices.

The devices are the MBR6535/45 and MBR8035/45. The first two digits in the type number indicate the current rating in amperes when T_C is 120 °C at the rated V_R (blocking voltage). The second pair of digits shows the voltage rating. For example, the MBR6535 is a 65-amp, 35-volt device. Similarly, the MBR8045 is a 45-volt device rated at 80 amps. These devices are ideally suited for use as rectifiers in low-voltage, high-frequency inverters and for service as free-wheeling diodes and polarity-protection diodes.

Chapter 16

Thermometer Circuits

In the last chapter, we looked at the REF-01 and the REF-02 from PMI (1500 Space Park Drive, Santa Clara, CA 95050). Accurate thermometers, reading directly in °C or °F can be built using either device, a few external parts, and a voltmeter. Scales can be based on references such as 10 mV/°C, 100 mV/°C, or 10 mV/°F, and direct voltage-readings such as −0.55 volts at −55°C, 0 volts at 0°C, and +1.25 volts at +125°C.

Figure 16-1 shows a simple thermometer circuit that is built around the REF-02. The only additional components are an op amp and three resistors. Those components, and the 5-volt output of the REF-02, scale and amplify the temperature-dependent V_{TEMP} to produce a usable V_{OUT} at the opamps output. That voltage output is determined from the equation:

$$V_{OUT} = \left(1 + \frac{R_c}{R_a \| R_b}\right) V_{TEMP} - \frac{R_c}{R_a} (V_{REF})$$

where $R_a \| R_b$ is the parallel combination of R_a and R_b. The term enclosed in the parentheses is the circuit gain when V_{REF} equals zero. The term to the right of the parentheses is the circuit gain when V_{TEMP} equals zero. The slope, S, of the temperature-versus-output curve (TCV_{OUT}) is a factor of the value of the three

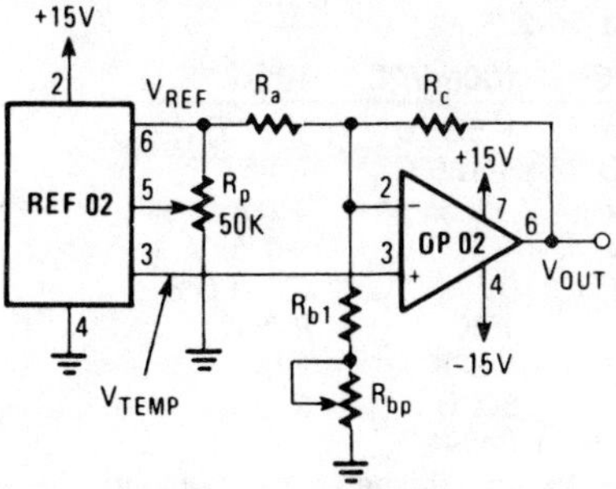

Fig. 16-1.

resistors and is determined from:

$$S = 2.1\text{mV}/°\text{C}\left(1 + \frac{R_c}{R_a \| R_b}\right)$$

From the above equation, we can see that the ratio of R_c to $R_a \| R_b$ determines the slope of V_{OUT} and the ratio of R_c to R_a and V_{REF} determines the level of V_{OUT} at 25 °C. Table 16-1 shows typical scaling-ratios for different values of V_{OUT}.

Table 16-1.

V_{OUT} at 25°C (77°F)	TCV_{OUT} (Slope)	R_c/R_a	$R_c/R_a \| R_b$
250mV	10mV/°C	0.55	3.76
2.5V	100mV/°C	5.50	46.6
770mV	10mV/°F	0.926	7.57

V_{REF} = 5.000V, V_{TEMP} = 630mV at 25°C, TCV_{TEMP} = 2.1mV/°C

A practical circuit—ready to drive a suitably calibrated voltmeter—is shown in Fig. 16-2. The 50K potentiometer, R_p, is used to adjust V_{REF} over a range of 5 volts ±300 mV to compensate for any error introduced by the circuit, and to permit precise calibration using 1% resistors. Resistor values are given in Table 16-2.

Detailed calibration instructions and performance specifications

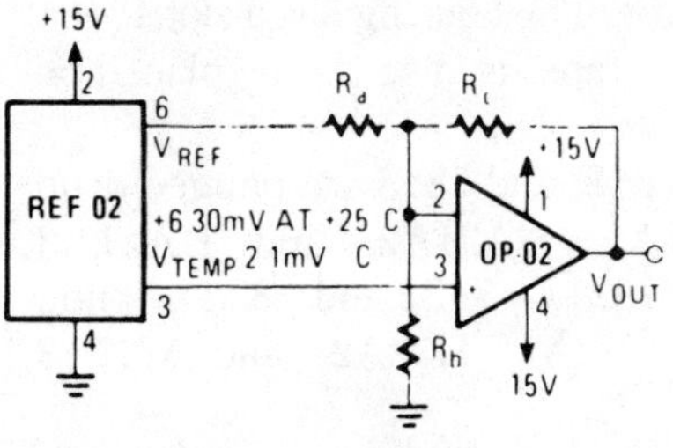

Fig. 16-2.

Table 16-2.

TCV_{OUT} slope (S)	10mV/°C	100mV/°C	10mV/°F
Temperature range	−55° to +125°C	−55° to +125°C	−67°F to +257°F
Output voltage range	−0.55V to −1.25V	−5.5V to −12.5V*	−0.67V to +2.57V
Zero scale	0V at 0°C	0V at 0°C	0V at 0°F
R_a (±1% resistor)	9.09K	15K	8.25K
R_{b1} (±1% resistor)	1.5K	1.82K	1.0K
R_{bp} (potentiometer)	200Ω	500Ω	200Ω
R_c (±1% resistor)	5.11K	84.5K	7.5K

*For 125°C operation, the op amp output must be able to swing to +12.5V; increase V_{IN} to +18V from +15V if this is a problem.

for the REF-02 when it is used as a temperature transducer can be found in PMI'S Application Note AN-18, "Temperature Applications of the REF-02."

LOW-DRAIN VOLTAGE-REGULATOR ICS

The new Intersil (10710 Tantau Ave., Cupertino, CA 95014) ICL7663 and ICL7664 monolithic voltage regulators feature a maximum quiescent current (I_Q) of only 4μA—1000 times better than a typical bipolar voltage-regulator. With a 1-mA load, the V_{IN}/V_{OUT} differential is only 50 mV. The low I_Q and low voltage differential make possible the use of smaller batteries and possibly allow longer battery life.

The devices handle operating loads ranging from 1μA to 40 mA at from 1.6 to 16 volts. Output voltage is set by a two-resistor voltage divider—the only external components required. The ICL7663 is for positive voltages and the ICL7664 for negative voltages.

SLOTTED OPTOCOUPLERS

A series of six new slotted optocoupler/interrupter devices has been introduced by Motorola that are drop-in replacements for the GE H21 and H22 series. The devices, encapsulated in a molded plastic housing, consist of a gallium-arsenide infrared-emitting diode facing a silicon NPN phototransistor. The housing has a slot so that an opaque material such as a card, tape, or disc can be placed between the emitter and detector.

The Motorola MOC7811, −7812, and 7813 are flanged drop-in replacements for the H21A1, HA21A2, and HA21A3, respectively. Similarly, the MOC7821, −7822 and 7823 are non-flanged replacements for the H22A1, H22A2, and H22A3, respectively.

Chapter 17

Power Transistor Driver/Amplifier

Whenever hi-fi experimenters or audio engineers begin a high-power audio amplifier, they are immediately faced with the question of how to interface and op amp or low-level discrete voltage-amplifier devices to the power amplifier. The 741 or a similar op amp—operating from power supplies of ±6 to ±12 volts and delivering a maximum of around 5 mA—cannot drive power transistors and Darlington configured devices. In a 50-watt RMS amplifier operating from a ±35-volt supply, those devices require 50 mA or more of drive.

Most approaches to the problem have used two or more discrete driver stages—each with its own power-supply requirements, SOA (*S*afe *O*perating *A*rea) protection, and short-circuit protection.

Intersil has taken another approach—they have developed a dedicated IC that is the total solution to the problem of driving almost all power transistors with breakdown voltages up to 70 volts. The device—the ICL8063—is a monolithic power-transistor driver and amplifier. It is intended primarily for complementary-symmetry outputs in an audio amplifier and as a driver for linear or rotary actuators, and servo and stepping motors. It is compatible with most op amps and dedicated devices such as preamps and compandors; taking output levels in the order of ±11 volts and boosting them to ±30 volts at 100 mA to drive power transistors. For example, Intersil used 2N3055 (NPN) and 2N3791 (PNP) as the output transistors in their data-sheet circuits. The ICL8063 includes built-in

±13-volt regulated outputs to power op amps or other external devices. Therefore, only ±30-volt supplies are needed for a complete power amplifier.

Using the ICL3068, we can build a power amplifier delivering ±2 amps at ±25 volts with only three additional discrete devices (a pre-driver and two power transistors) and as few as eight passive components. The slew rate of the power amplifier is the same as that of the 741 pre-driver by itself; except that the output current can slew up to 2 amps at 1V/μs. Other factors such as common-mode rejection ratio (CMRR), input current, voltage offset and power-supply rejection ratio (PSRR) are also the same as for the 741 op amp. Typically three 1000-pF (.001 μF) compensating capacitors are used to ensure good stability down to unity gain. The circuit will drive a 1000-pF load (as might be represented by 30 feet of RG-58 coaxial cable) in line-drive application, without problems. Quiescent current is only 30 mA from a ±30-volt power supply.

A $20-PER-CHANNEL 50-WATT AMPLIFIER

Figure 17-1 is the schematic of a power amplifier using the ICL8063 to drive 2N3055 and 2N3791 power transistors to 50 watts

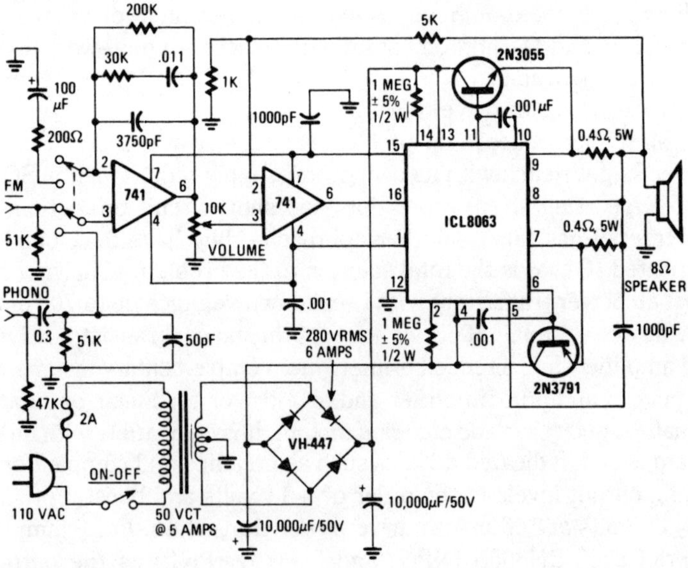

Fig. 17-1.

into an 8-ohm load. (The pinout of the ICL8063 is shown in Fig. 17-2.) The first 741 is a preamp for FM tuner and phonograph inputs. The phono input has RIAA (*R*ecording *I*ndustry *A*ssociation of *A*merica) equalization. The second 741 is a pre-driver for the ICL8063. The complementary-symmetry-output transistor stage delivers 56 volts P-P (50 watts RMS) into an 8-ohm speaker. Distortion is less than 0.1% up to about 100 Hz and increases to about 1% at 20 kHz.

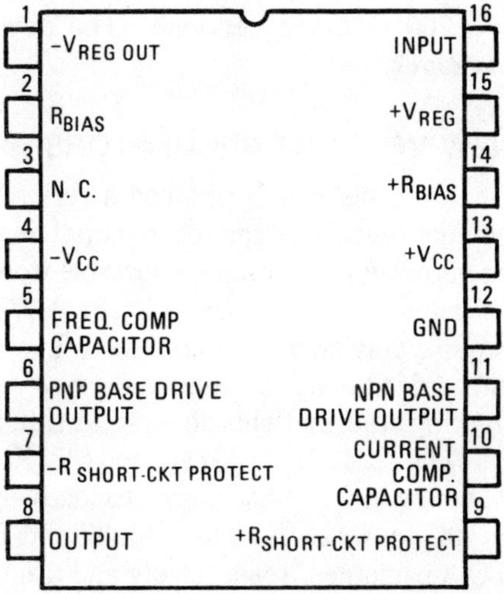

Fig. 17-2.

The 0.4-ohm resistors limit the maximum output current that can be drawn. The 1-megohm biasing resistors (between pins 2 and 4 and 13 and 14) are based on $V_{CC} = \pm 30$ volts and guarantee adequate performance when driving DC motors, programmable power supplies, and power DAC's. You can decrease V_{CC} from ± 30 to ± 5 volts in 5-volt steps by using 1 megohm, 680K, 500K, 300K, 150K, and 62K biasing resistors.

When selecting the output transistors for the amplifier, make sure that their beta (hfe) does not exceed 150 at $I_C = 20$ mA and $V_{CE} = 30$ mV. The output terminal can be shorted to ground for an indefinite period as long as the transistors have adequate heat sinks.

DUAL HIGH-SPEED RECTIFIERS

The RUR-D1610, -D1615, -D1620 series is a new family of RCA ultra-high-speed dual-chip rectifiers intended as output rectifiers and fly-wheel diodes in high-frequency pulse-width-modulated power supplies and switching regulators. The devices feature a current-carrying capacity of 16 amps per diode and a recovery time of less than 35 ns. Maximum forward voltage drop (at 25 °C and full rated current) is 0.95V.

The low stored-charge and fast recovery of the RUR series minimizes electrical noise and, in many circuits, reduces the turn-on power dissipation of associated switching transistors. Breakdown voltages for the RUR-D1610, -D1615, and -D1620 are 100, 150, and 200 volts, respectively. The devices are in steel TO-204M packages.

NEW VOLTAGE REGULATORS

Motorola has introduced a series of three-terminal negative-voltage regulators capable of supplying in excess of 1.5 amps over an output-voltage range adjustable from – 1.2 to – 37 volts. These voltage regulators—the LM137/237/337—are easy to use and require only two external resistors to set the desired output voltage. Added features found in the new series include internal current limiting, thermal shutdown, and safe-area compensation.

The LM337T, 337H, and 337K are packaged in TO-220, TO-39, and TO-3 housings, respectively. Their temperature range is °0C to + 125 °C. The LM237 and LM237K are in TO-39 and TO-3 packages, respectively and are developed to operate in the – 25 °C to + 125 °C range. The LM137H and LM137K operate over a – 55 °C to + 150 °C temperature range and are in TO-39 to TO-3 packages, respectively.

MEMORY DESIGN KIT

The TMS4500-Kit from Texas Instruments is the firm's new *MOS Memory Design Kit* that comes with enough devices and supporting literature to permit you to design a 32K byte memory system. The kit contains four TMS4416's—the firm's newest DRAM (*D*ynamic *R*andom *A*ccess *M*emory)—and the new TMS4500A DRAM controller featuring an address multiplexer, refresh counter, timing and control circuits, and logic for microprocessor access and memory-refresh sequences.

Chapter 18

High-Voltage Transistors

Even though we all know that high-voltage transistors are available, many of us still think of the transistor as a low-voltage device. And we would be hard-pressed to draw a circuit illustrating a simple application of a high-voltage transistor. That's not true of STI (Semiconductor Technology Inc.), one of the industry leaders in high-voltage-transistor technology. They manufacture a broad line of silicon high-voltage types. The low- and high-power devices have ratios ranging from 30mW to 200 watts and voltage ratings up to 2kV V_{CBO} (collector-base breakdown voltage with open emitter). Many of those are direct replacements for discontinued types that were made by Delco and Texas Instruments.

The circuits in Figs. 18-1, 18-2, and 18-3 (from an STI application note) illustrate the development of a high-voltage wide-band deflection amplifier for the CRT in an oscilloscope. Figure 18-1 is the circuit of a basic Class-A amplifier. It has a low input impedance and a voltage gain of over 300. A disadvantage of this circuit is that the Miller effect multiplies the collector-to-base or feedback capacitance by about the value of the gain. That very high effective input-capacitance loads the input circuit, reduces the high-frequency response, and increases the effective input-to-output coupling. The collector voltage is 300 volts with an idling or quiescent current of 10 mA, and the output voltage swing is 600 volts peak-to-peak.

Figure 18-2 shows how—through the addition of a low-voltage,

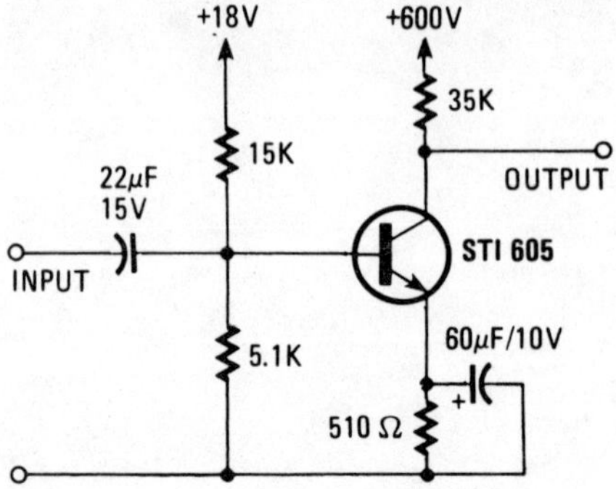

Fig. 18-1.

high beta, NPN silicon transistor such as the 2N2219-we can improve performance by greatly minimizing the Miller effect and almost completely eliminating input circuit loading. The transistors are stacked in what is often called a cascode configuration with the 2N2219 as a common-emitter amplifier driving the emitter of the STI 605 that is operating in the grounded-base mode. The input signal is fed to the base of the 2N2219. The STI605 provides the necessary voltage gain while the 2N2219 provides current gain. The collector of the 2N2219 sees only a 1-volt peak-to-peak swing.

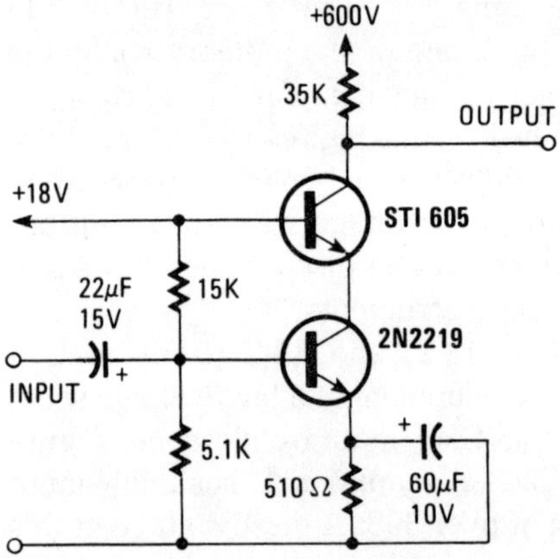

Fig. 18-2.

One of the advantages that is offered by this arrangement is that the STI 605 is operating under conditions that minimize the effects of beta changes and leakage currents, so the stability factor approaches one. That's accomplished by biasing the base of the STI 605 base with a constant voltage while emitter is biased by the 2N2219, which acts as a constant-current source.

Figure 18-3 shows how two stages of the circuit just described can be arranged as a differential amplifier delivering approximately 1200 volts peak-to-peak. The circuit can be used as a wide-band deflection amplifier in a scope. The third 2N2219 is a constant-current source for the differential amplifier.

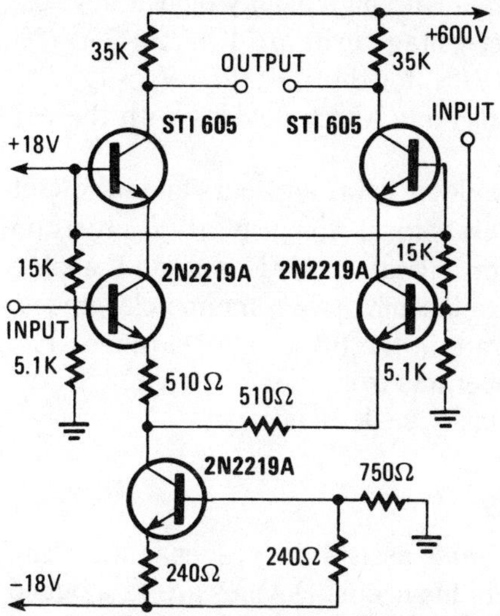

Fig. 18-3.

Taking a quick look at the characteristics of the STI 605, we see that its power dissipation at 25 °C is 2 watts; V_{CBO} is 600 volts; I_{CBO} (the collector-base cutoff current) at 450 volts is 4 μA; h_{FE} (the DC current gain in the common-emitter configuration) at I_C = 25 mA is 30; and V_{CEsat} (the collector-emitter saturation voltage) when I_C = 25 mA is 1.5 volts.

DRAM CONTROL APPLICATION MANUAL

A 42-page booklet, *TMS4500 Dynamic RAM Controller User's Manual* (SCG690A) describes the operation of the Texas Instruments TMS4500 dynamic memory controller and illustrates its applications in a microprocessor system. The device provides address multiplexing, cycle timing, and refreshing for DRAM's (*D*ynamic *R*andom-*A*ccess *M*emories). Chapter headings are "Functional Description," "Typical Implementation," "Design Criteria," and "Application Examples." The last chapter discusses uses of the TMS4500A with the 8085A, Z80, TMS-9900, MC68000, 8086, Z8001, and TMS-9995 microprocessors. It's a handy guide if you are designing or updating a small computer or microprocessor-based equipment.

BI-FET OP-AMP

The AD547 is a monolithic, FET-input op amp from Analog Devices. It offers the low-input-bias-current (typically 25 pA) advantage of the Bi-FET with the low-offset-voltage and low-drift characteristics usually associated with high-quality bipolar devices. Input offset voltage and offset-voltage drift are 1 mV, $\pm 5\ \mu V/°C$ for the AD547J; 0.5mV, $\pm 2\mu V/°C$ for the −K; 0.5mV, $\pm 1\mu V/°C$ for the −L, and 0.5 mV, $\pm 5\mu$Vmu/°C for devices with the SH suffix.

The low noise, high open-loop gain, and outstanding offset performance make the AD547 ideal for use as a precision instrumentation amplifier for signal conditioning and analog computations. The device is available in three commmercial grades. The J, K, and L grades operate in the 0° to +70 °C range. The SH grade (military version) operates from −55 °C to +125 °C. All grades are in TO-99 hermetically sealed cans.

DEDICATED IC CATALOG

Microelectronics Product Guide assists the reader in selecting the dedicated IC best suited to his needs. Devices are catalogued according to category under such headings as "ROM's," "Keyboard Encoders," "Character Generators," "EEPROM's/EAROM's," "Microcomputers and Microcomputer Development Systems," "Speech Synthesis," "TV Games," and "Radio and TV Control." Devices in each category are listed in a tabular form. In addition to part numbers, there are column headings for function, description, package type, supply voltage, and special features.

POWER SEMICONDUCTOR DATA MANUAL

Semiconductor Product Digest-1982-1983 (Catalog SPD-1) is a 25-page short-form listing of the characteristics of International Rectifier's power semiconductor devices. Devices are listed by construction and purpose so they can be quickly located. Groupings include *HEXFET* Power MOSFET's, Power Schottky Rectifiers, Bipolar Power Transistors and Darlingtons, Thyristors; General-Purpose, Fast-Recovery, and High-Voltage Silicon Rectifiers; Diode Bridges, Protective Devices, and Opto-electronic Devices. An alphanumeric listing greatly simplifies finding the location of a particular device in the manual.

Chapter 19

An Interesting "Melody-Maker" IC

We have recently come across several dedicated ICs that are used in unusual and/or interesting applications. The one we'll look at this month is the MP1173A-NI. It's a thirty-tune melody microprocessor from SRJ International (1936 Hillman Avenue, Belmont, CA 94003). The 4-bit microprocessor is a P-channel MOS device with on-chip ROM, RAM, and ALU. It's programmed to play thirty different melodies electronically.

The 28-pin DIP device can be used for many applications. For example, it can be used as a door chime, a replacement for your telephone bell, an intercom annunciator, or a music box. In your car, it could be used as a musical horn or for warning signals (door ajar, low fuel, etc.). A few inexpensive discrete components can be added for melody selection, automatic melody-sequencing, and additional power-amplification (when needed). Briefly, the features offered by the MP1173 are:

1. It plays thirty different melodies.
2. Sixteen tunes can be played automatically in sequence.
3. Melody selection can be controlled by two six-position rotary switches.
4. Its tone and volume are variable.
5. Tune speed is variable.
6. A 50-60-ohm speaker can be directly driven.
7. It can be powered by two 9-volt batteries.

8. Its scale is chromatic.
9. Tempo is programmed for each melody.

Figure 19-1 shows how the IC can be used for a door-chime (or for a similar application). The optional CD4015 CMOS 4-bit shift register (IC2) is added so that, with S1 open (in the AUTO position) the melodies are automatically sequenced as S4 is pressed and then released. The door chime can be operated from two 9-volt batteries or from rectified AC from a 12-18-volt bell transformer. Figure 19-2 shows how a general-purpose power transistor—fed from output pins 12 through 17—can be used to obtain higher volume or to drive a low-impedance speaker.

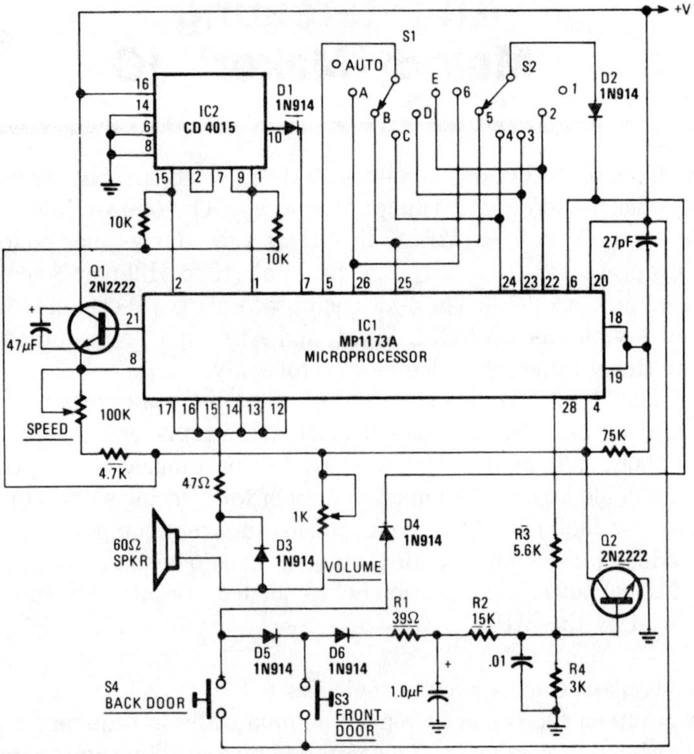

Fig. 19-1.

In its standby mode, the MP1173 is turned on but consumes very little power. The CD4015C (IC2) has a quiescent current drain ranging from 1 μA (typically) to a maximum of 100 μA when V_{DD}

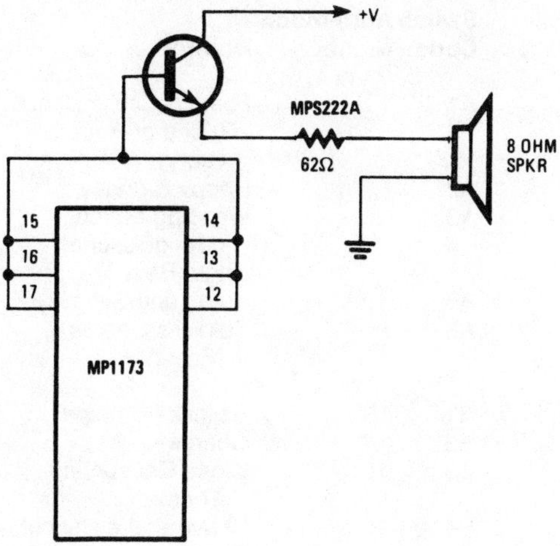

Fig. 19-2.

is 10 volts. Thus, for the circuit shown, the battery's life should approximate its "shelf life".

HOW IT WORKS

When either momentary switch (S3 or S4—the back- and front-door bell buttons) is pressed, current through R1, R2 and the diode(s) causes Q2 to saturate. That, in effect, grounds pin 4 (IC1's V_{DD} terminal) and turns on the microprocessor. Output pin 28 goes high and the base of Q2 is now held on by current flow through R3 and R4. This keeps IC1 turned on after momentary switch S3 or S4 is released. At the end of the melody selected by S1 and S2—or the next tune in sequence if the chimes are set in the automatic-sequence mode—pin 28 of the MP1173 goes low and Q2 turns off.

If S3 is held down (closed) the tune selected by S1 and S2 repeats or the sixteen melodies will play in sequence if the chimes are set for automatic sequencing. A momentary closure of S4—the backdoor button—connects pin 6 to the positive supply voltage through diode D4 and the chimes will play the single tune "Oranges and Lemons." Holding S4 down causes the melody to repeat.

Switches S1 and S2 select the thirty different melodies according to the letter/number combinations listed in Table 19-1. The table also lists the sequence of the 16 melodies that are played when S1 is in the AUTO position.

SRJ also sells the *Phone-Tunes*, which is an FCC approved telephone-bell replacement that connects easily to the telephone

Table 19-1.

Switch Code	Auto-mode Number	Melody
A1		Pomp & Circumstance (Land of Hope & Glory)
A2		Happy Birthday
A3		Wedding March (Mendelssohn)
A4		Jingle Bells
A5		Auld Lang Syne
A6		Soldier's Chorus (Faust)
B1	13	Sailor's Hornpipe
B2	14	Charge!
B3	15	Close Encounters Theme
B4	16	Shave and a Haircut
B5		Rule Brittania
B6		O Canada!
C1	9	Colonel Bogey March
C2	10	Westminster Chimes
C3	11	Mexican Hat Dance
C4	12	Twinkle, Twinkle Little Star
C5		Deutschlandlied
C6		God Save The Queen (My Country 'Tis of Thee)
D1	5	William Tell Overture
D2	6	Greensleeves
D3	7	Lorelei
D4	8	Eyes of Texas (I've Been Working on the Railroad)
D5		Star Spangled Banner
D6		*Oranges & Lemons
E1	1	Mozart Sonata
E2	2	Blue Danube Waltz (Strauss)
E3	3	Beethoven's 5th Symphony
E4	4	Bach Toccata in D Minor
E5		La Marseillaise
E6		Wilhelmus

*Back door tune:

system. It, of course, uses the MP1173A-NI IC. Thirty familiar tunes can be selected and different tunes can be used to identify different phones.

MIKE ON A CHIP

Honeywell has just developed a new type of microphone using integrated-circuit technology that consists of a thin film of zinc oxide on a silicon substrate that is direct-coupled to the gate of a PMOS FET amplifier. The sensing element of the new mike is passive; the FET amplifier dissipates less than 40 mW so it can operate for many months before the batteries need to be replaced.

Zinc oxide, like piezoelectric crystals and ceramics, produces electric charges when subjected to stress. The new mikes have response down to 0.1 Hz while the response of most ceramic mikes is well down at 20 Hz.

The integrated-circuit sensors are lighter and smaller than their ceramic counterparts. They are one-quarter inch square while ceramic mike elements are often one-quarter inch thick and one-half inch in diameter. Their reliability is greater because they are solid state—no parts to cement or solder as with ceramics.

The new mike is expected to have many applications. For example, it is expected to reduce hearing-aid size.

LOGIC OPTOCOUPLERS

Motorola has announced a series of infrared optocouplers with Schmitt-trigger outputs for coupling digital logic circuits in situations that require a high degree of electrical isolation between the control circuits and the controlled equipment. The digital output of the couplers eliminates the need for comparators or other wave-shaping circuitry between computer terminals and peripheral equipment. The infrared optocouplers also have many applications in digital control of power supplies, motors, and servos.

The MOC5007, H11L1, MOC5008, MOC5009, and H11L2 optocoupler/isolators feature a V_{ISO} (isolation voltage) of 7500-volts AC minimum, and guaranteed switching times (t_{on}, t_{off}) less than 4 μs. The LED trigger current is specified as 1.6 mA for the first two devices, 4 mA for the MOC5008, and 10 mA for the MOC5009 and H11L2. The rise and fall times for the output of the Schmitt-trigger waveform is specified as 0.1 μs.

The H11L1, and H11L2 are equivalents of devices introduced earlier by General Electric.

Chapter 20

IC Metal-Sensing Devices

Recent developments in the field of electronics technology have given us some interesting semiconductors. We'll take a look at some of those devices and their applications.

First let's look at two bipolar ICs from Cherry Semiconductor—the CS191 and CS209. Those ICs are metal-sensing devices and can be used in a wide variety of applications—including electronic ignitions and metal detectors. Figure 20-1 is a block diagram of the CS191 and Fig. 20-2 is the block diagram of the CS209.

Each IC contains a voltage regulator, oscillator, demodulator, and a level detector. The CS209, which we will look at, has only two high-level outputs (pins 4 and 5) for external loads. The output of pin 4 is normally open and pin 5 is normally closed (high). (The CS191 also has two low-level outputs with internal loads.) The saturation voltage for the high-level outputs is 0.2 volt at 124 mA (0.03 volt at 2 mA for low-level outputs). Both ICs require a supply voltage between 4 and 24 volts, and draw about 4.5 mA when operating from a 4-volt supply.

The internal oscillator, together with an external L-C network, provides an output signal whose amplitude is highly dependent on the Q of the external L-C tank circuit. To sustain the oscillations when the Q is very low, a variable low-level feedback signal is developed. Both ICs have transient suppressors to protect their internal circuitry against transients (spikes) that might develop in the tank circuit.

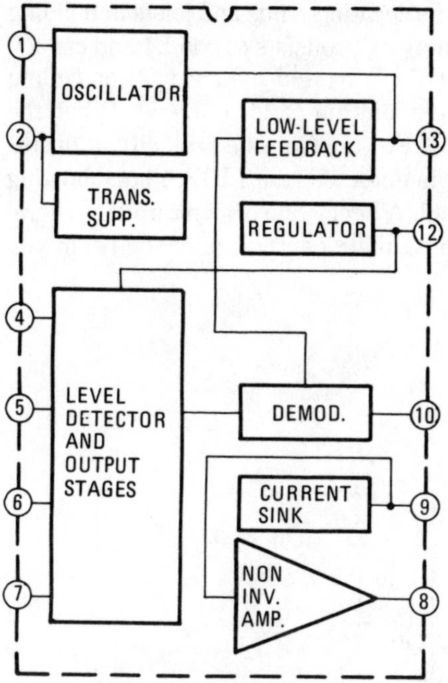

Fig. 20-1.

The demodulator rectifies the oscillator's output and the resulting DC voltage is fed to the level detector, where it's compared to an internal reference to produce an output signal.

BUILD A STUD DETECTOR

The diagram in Fig. 20-3 shows how the CS209 can be used

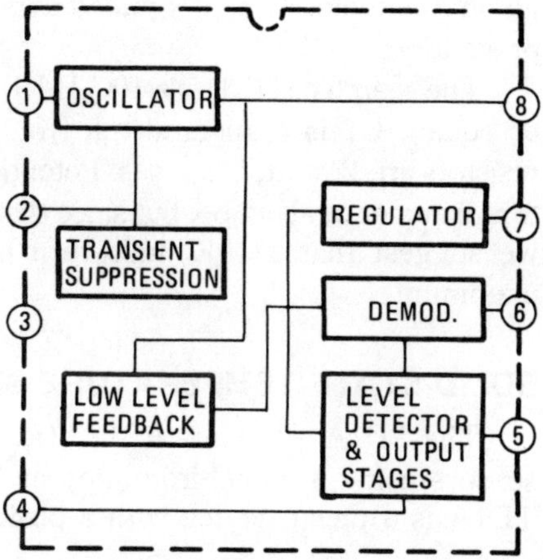

Fig. 20-2.

to build a pocket-size device for locating studs and joists in building walls and ceilings. The L-C network consists of coil L1 and capacitor C1. The "search" coil (L1) is passed over a wall or ceiling surface to locate a beam by pinpointing nails or screws. When the coil comes close to a nail or screw, the Q of the tank circuit drops. Whenever that change in Q is detected, the LED lights, showing the location of the screw or nail. A pencil line drawn through several nail positions shows the approximate center line of the beam you are looking for.

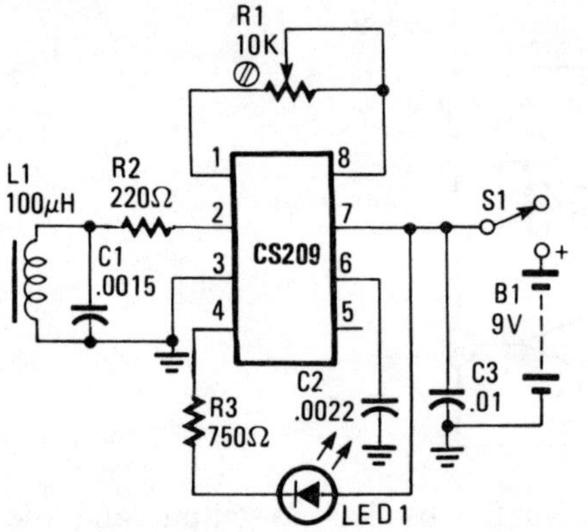

Fig. 20-3.

The sensitivity of the device is adjusted using potentiometer R1. To accomplish that, set R1 so that the LED lights when the search coil is well clear of any metal objects. Then slowly adjust the trimmer in the other direction until the LED just goes out. The LED should light when the device is brought close to a nail and go out immediately as it's pulled away. If not, repeat adjustment procedure.

The search coil (L1) is a 100μH RF choke (Radio Shack 273-102 or equal), C1 is a silvered mica or polystyrene capacitor. The resistors are ¼-watt, 5% types. Potentiometer R1 should be a multi-turn trimmer of about 6K but since that value isn't readily available, we suggest that a 10K multi-turn potentiometer be used as a maximum.

SOLID-STATE TEMPERATURE SENSOR

The TD1A is the first in the Micro Switch line of temperature sensors using a nickel-iron alloy and thin film technology. The TD1A is a linear device with a positive temperature coefficient (sensor resistance increases with temperature). Its applications include room, duct and refrigerant monitoring, motor-overload protection, electronic-equipment overheat warning, and cooking-temperature settings for appliances.

The IC is constructed of nickel-iron alloy deposited on a 0.040-inch square silicon chip and is laser-trimmed to provide a stable room temperature resistance accurate to 0.4 °C. It is then mounted on a 0.2-inch ceramic substrate and epoxy encapsulated for protection.

The TD1A has a higher resistance than that of platinum sensors—thus, making it well suited for use in low-power circuitry and for minimizing errors caused by voltage drops between the sensor and control circuitry.

Its operating range is from -40 °C to $+150$ °C and its resistance at 20 °C is 2000 ohms. Its tolerance at 20 °C is $\pm 0.4°$ and ± 1.0 °C over the entire range. Its recommended excitation is 1 mA and it has a thermal time constant of 4 seconds maximum.

LOW-CURRENT LEDS

Hewlett-Packard has announced the latest in its series of low-current LEDs. Those LEDs use only 2 mA and are five times brighter than standard ones that draw 10 mA. Typical CMOS and low-power TTL circuitry provides enough current to maintain high brightness without external drivers. The HLMP-4700, HLMP-1700, and HLMP-7000 are red LEDs. The HLMP-4719, HLMP-1719 and HLMP-7019 are yellow. The HLMP-4700 series come in T-1¾ (5mm) packages and the -1700s in T-1 (3 mm) packages, and the -7000s are subminiature devices.

TRANSISTOR REPLACEMENT GUIDES

A four-page booklet, *Direct Replacement for Texas Instruments Transistors,* lists STI direct equivalents for 540 metal-can small-signal and power transistors discontinued by Texas Instruments.

Also available from STI are cross-references and data sheets of pertinent characteristics of replacement high-voltage power transistors in the DTS series of devices that was discontinued by Delco.

ISOLATED-FEEDBACK GENERATOR

The UC1901 series ICs from Unitrode features an AM-carrier system that replaces the visible or infrared light path in optocoupler/isolators. An internal RF oscillator and amplitude modulator develop the signal that's coupled across the voltage-isolation boundary by a small RF transformer. The oscillator is usable up to 5 MHz. An external clock can be substituted for the oscillator to synchronize the device to a system clock or to the frequency of a switching power supply.

As an added feature—a status monitor that provides an active-low output when the sensed error voltage is within ±10% of the 1.5-volt precision internal reference. The UC1901 operates on voltages between 4.5 and 40-volts. It's available in 14 pin plastic and ceramic DIP's—in military, commercial, and industrial versions.

Chapter 21

Telephone Ringer ICs

For years, many of us have found the strident tones of a telephone bell annoying and have wished for something with a more pleasant tone. Others have wanted an extension telephone ringer in the garage, workshop, or similar location, but never got around to having one installed by the telephone company. Now, Motorola has introduced the MC34012 telephone bell ringer IC, a simple solid-state device that can be used as a solution to both problems. A pinout of that device is shown in Fig. 21-1.

Let's see what the pin functions are: Pin 1 (RG) is the negative output of the internal diode bridge. Pins 2 and 3 (AC_1 and AC_2) are the input terminals for the AC telephone-line ringing signal. Pin 4 (RO) is the ringer output terminal. Pin 5 (RI) is the positive supply terminal for the oscillator, frequency divider, and buffer/amplifier circuits. Pin 6 (RS) connects to an external sense resistor, pin 7 (RF) connects to a filter capacitor, and pin 8 (RC) connects to an external R-C circuit that controls the frequency of the internal oscillator.

The MC34012, a monolithic I^2L linear device, is housed in an 8-pin plastic DIP. That IC is used to drive a piezoelectric ceramic sound transducer to produce a pleasant warbling tone at a suitably high audio level. It is available in three output-frequency ranges, and those ranges are indicated by the suffixes −1, −2 and −3 that identify the 1-kHz, 2-kHz and 500-Hz devices, respectively.

The MC34012-1 has a base frequency of 4 kHz and produces

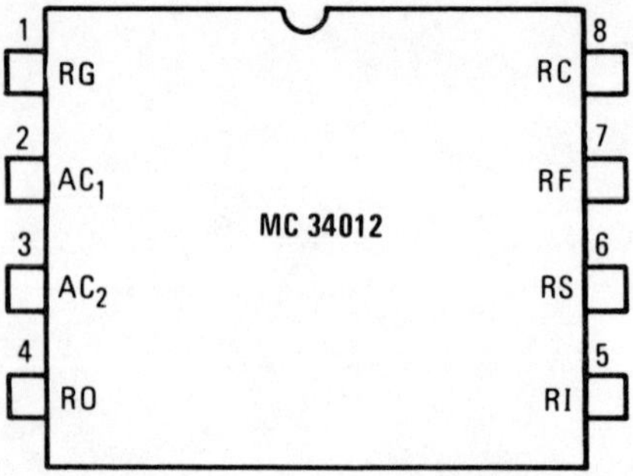

Fig. 21-1.

800-Hz and 1000-Hz tones. Similarly, the MC34012-2 oscillator operates at 8 kHz and generates 1600 and 2000-Hz tones. The MC34012-3 has a 2-kHz oscillator and produces a 400- and 500-Hz tone. The warble frequency of the devices is 13 Hz. The tone ringer output can source or sink 20 mA with an output-voltage swing of 20 volts peak-to-peak.

The input impedance presented to the telephone line by the devices meets Bell and EIA requirements. Typical performance characteristics of the three devices are given in a table available from the manufacturer. Now, let's look at how those devices work.

HOW THE RINGER WORKS

Figure 21-2 shows the functional block diagram of the MC34012. It consists of a relaxation oscillator and two frequency dividers that provide the high- and low-frequency segments of the warble tone, as well as the 12.5-Hz warble rate. The IC is connected across the telephone line through the series network consisting of R1 and C1. The line-input resistor, R1, controls the ringer's input impedance as seen by the telephone line. That resistor also limits potentially harmful current surges caused by line transients, and influences the ringing-threshold voltage. The value of R1 can range from 2000 to 10,000 ohms. Capacitor C1 provides AC coupling between the line and the IC, and also controls the ringer's input impedance at low frequencies. The value of that capacitor can range from 0.4 to 2.0 μF.

The base frequency of the relaxation oscillator, which can range between 1 kHz and 10 kHz, is determined by the values of R2 and C2. the capacitance value of C2 may range between 400 pF and 2000 pF. Frequency-determining resistor R2 may be a selected value between 150 and 300 kilohms. The frequency of the warble

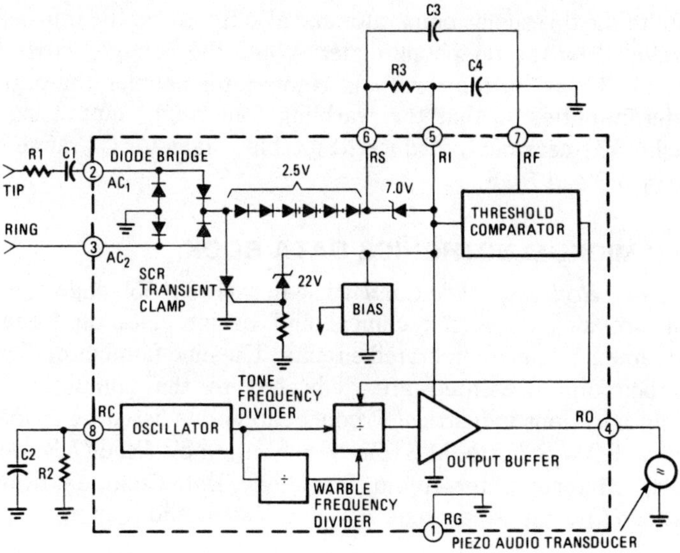

Fig. 21-2.

tone at the ringer output terminal (pin 4) switches between $f_0/4$ and $f_0/5$ at a 12.5-Hz rate.

The DC voltage needed to power the IC is obtained by rectifying and filtering the 20-Hz AC ring-signal voltage (60 to 90 volts RMS). That DC voltage supplies power to the tone generator and the circuits needed to drive the ceramic transducer. The ringing-signal voltage is rectified by the full-wave bridge and develops currents along two paths. In one path, an internal Zener diode between pin 5 and ground provides 22 volts to operate the oscillator, frequency dividers, and buffer/amplifier output stage. Capacitor C4 (1.0 to 10.0 μF) filters the supply voltage for the tone-generator circuits and provides an AC-current path to ground for the 10-volt RMS ringer signal when the phone ringer is not driving a load. Potentially harmful line transients trigger the SCR when the Zener regulator current exceeds about 50 mA. When the SCR fires, it diverts current from the shunt regulator and reduces the power dissipated within the IC.

In the second current path the rectified ring-signal voltage produces a current flow through R3, the input-current sense-resistor, to control the ringing-threshold voltage. The design range for R3 is 800 to 2000 ohms; increasing that resistor decreases the ring-start voltage. Capacitor C3 filters the voltage across R3 at the input to the threshold comparator and also filters out the transients developed by the telephone dialer. When the voltage across C3 exceeds 1.7 volts, the threshold comparator enables the output buffer/amplifier so that the warbling tone-ringer output signal reaches the ceramic transducer. Suitable values for C3 range between 0.5 and 5 μF.

NEW MICROELECTRONICS DATA BOOK

Microelectronics Data Catalog is a new (1983) 559-page edition that provides detailed technical and design data on General Instrument's line of integrated circuits. The nine families of digital microcircuits presented are dedicated to the computer and communications industries. Product categories listed by chapter, include ROM'S, EAROM'S (*E*lectrically *A*lterable *R*ead-*O*nly *M*emories), Microcomputer, Audio, Telephony, Data Communications, ULA's (*U*ncommitted *L*ogic *A*rrays), Video, and Tuning.

Chapter 22

Making True RMS Measurements

The accuracy of measured RMS values of AC voltages has always been questionable. That's because the shape of the AC wave being measured can drastically affect the readings of many voltmeters. In most multimeters and AC panel-meters, the AC wave is rectified to develop a DC voltage that is proportional to the RMS value. For example: two popular analog multimeters specify their AC voltage accuracy as 5% of full-scale value when measuring a 60-Hz sinewave. Accuracy "goes out the window" if the input signal's shape is nonsymmetrical or if the signal approaches a triangle or pulsed waveform.

Semiconductor manufacturers have developed monolithic RMS-to-DC converters that permit AC voltages to be measured with maximum errors that are a small fraction of one percent. A recent addition to this category of devices is the AD637 from Analog Devices.

That IC is a monolithic, high-accuracy RMS-to-DC converter that can be used to compute the true RMS value of any complex AC waveform. In addition, the device computes the square, mean-square, and absolute values of complex AC (or AC superimposed on DC) input waveforms. Its circuit features a wide bandwidth: 8 MHz when the input is 2 volts RMS or above, and 600 kHz with a 100-millivolt RMS input. The circuit includes built-in compensation for *crest factor* and handles signals with crest factors of up to 10 with less than a 1% increase in error. Crest factor, CF, is the ratio of a signal's peak amplitude to its RMS value: CF = $(V_{PEAK})/(V_{RMS})$. Sine and triangle waves have low crest factors that do not exceed 2. Pulse trains with low duty-cycles have much higher crest factors. For example, a rectangular pulse series with a 1% duty cycle has a crest factor of 10.

A voltage equal to the logarithm of the RMS output signal is brought out to a separate pin on the AD637 so that the user can make direct measurements in decibels. The dB-measurement range is 60 dB; the 0-dB voltage-reference level can be set to between 0.1 volt and 2.0 volts RMS by providing an external reference current.

HOW THE CIRCUIT FUNCTIONS

The only external component required for RMS-to-DC conversion is a capacitor that controls the averaging-time period and determines the low-frequency accuracy, ripple level, and settling time.

Figure 22-1 is a functional diagram of the IC. We can think of it as being made up of 4 main sections. The first is an absolute-value circuit (active rectifier) made up of A1.A2. and their associated components. The second is a square/divider made up of A3 and Q1-Q5. Third is a filter/amplifier, A4, and fourth is buffer/amplifier A5. (The buffer amplifier can be used either as an input buffer or in an active-filter configuration; it is a user option.)

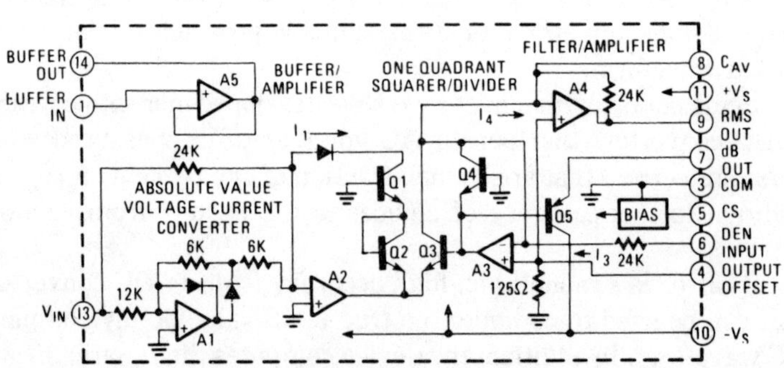

Fig. 22-1.

The input voltage, V_{in}, is rectified by active rectifier A1-A2 to produce a unipolar current, I_1. That current drives one input to the squarer/divider which has the transfer function: $I_4 = I_1^2/I_3$.

The output current of the squarer/divider, I_4, drives A4 which, along with the external averaging capacitor connected to pin 9, forms a lowpass filter. If we make the R-C time constant of the filter much longer than the longest period of the input signal. A4's output will be proportional to the average value of I_4. The output is taken back to A3 to provide current I_3, the denominator of the transfer-function equation above. See the standard RMS conversion circuit in Fig. 22-2. Current I_3 equals the average value of I_4 and is returned to the squarer/divider to complete the operation: I_4 = Avg $[I_1/I_4]$ = I_{IRMS} and V_{out} = V_{in} RMS.

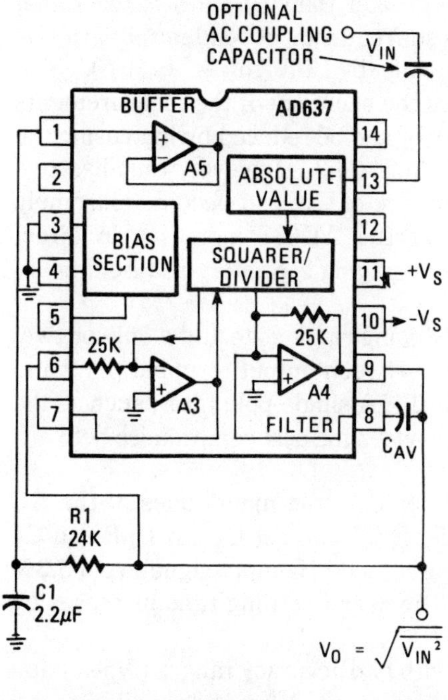

Fig. 22-2.

THE AVERAGING TIME CONSTANT

The AD637 can handle both DC and AC input voltages. Its DC output will track exactly the absolute value of a DC input and will approach the true RMS value of an AC input. The deviation from the true RMS value is the averaging error caused by AC ripples riding on the DC output signal. That is a factor of the input signal's frequency and the averaging time constant, τ, (25 ms/μF $\times$ C_{AV}) and is defined as the peak value of the AC ripple component plus the DC error. This is illustrated in Fig. 22-3.

The peak value of the AC ripple as a percentage of the output

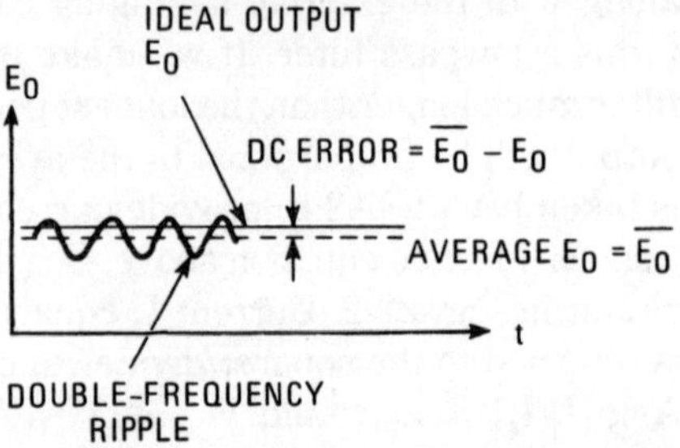

Fig. 22-3.

signal is $50/6.3\tau f$, where τ is greater than $1/f$ and f is the signal frequency. The DC error is also frequency dependent; it varies as the percentage of the reading by $1/(0.16 + 6.4\tau^2 f^2)$. The magnitude of ripple decreases the accuracy of the measurements being made. The amount of error can be reduced by increasing the value of the averaging capacitor, C_{AV}). However, that has two disadvantages: 1) The capacitance of C_{AV} can become extremely high. 2) The settling time of the AD637 increases in direct proportion to the value of the averaging capacitor. Setting time T_S equals 115 ms/μF $\times$ C_{AV}.

A preferred method of reducing ripple is to add a one- or two-pole filter network connected between output terminal pin 9 and the buffer amplifier input, pin 14. A single-pole filter, such as the R1-C1 network in Fig. 22-2, provides the best compromise between ripple and settling time.

When C1 is made 3.3 times C_{AV}, the magnitudes of the AC and DC errors are equal at 50 Hz. If we set C_{AV} to 1 μF and C1 to 3.3 μF, we reduce the ripple for a 60-Hz input signal from 5.3% (when C_{AV} is used alone) to 0.15% and settling time increases by not more than three times.

The AD637 is available with two accuracy ranges (types J and K) for a 0 to + 70 °C temperature range. A type S is available with an operating-temperature range of − 55 °C to + 125 °C. All devices are packages in ceramic 14-pin DIP's.

Chapter 23

Stereo-Sound IC

If you've grown accustomed to the big spatial sound of hi-fi stereo equipment, you probably find the monaural sounds of AM radios and TV sets somewhat less than enjoyable. With that thought in mind, Signetics has introduced the TDA3810, a spatial, stereo, and pseudo-stereo sound circuit. That audio-processor IC converts monaural-audio signals into realistic "pseudo-stereo" sound. It can also enhance the stereo effect when inserted in the audio-signal paths of FM-stereo radios where the apparent stereo separation is less than ideal because of limited speaker spacing (for example, in portables or in your car). It can also be added to most monaural TV and AM/FM radio circuits.

Figure 23-1 shows a schematic of a circuit using the TDA3810, along with a block diagram of the IC. The IC operates from a DC supply voltage ranging between 4.5 and 16.5 volts and draws a maximum of 12 mA. Its input impedance is typically 75,000 ohms. Minimum input-signal voltage (at pins 2 and 17) for the IC is 2-volts RMS rated at .2% total-harmonic-distortion (THD). Its output THD is rated at 0.1% (for frequencies between 40-16 kHz with an output voltage of 1 volt). The IC's average voltage gain is about 2.4 dB in the spatial mode, and unity otherwise. Its channel separation is typically 70 dB.

The TDA3810 also provides automatic switching between the mono and stereo modes. And it also has manually controlled logic switching between the mono and pseudo-stereo modes. (Mode

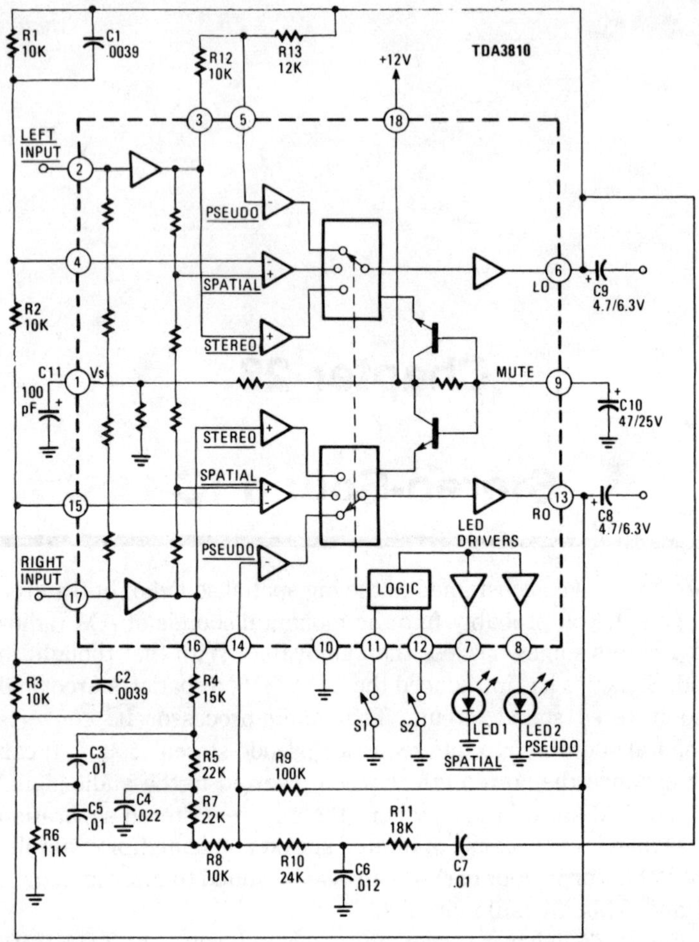

Fig. 23-1.

selection is determined by the settings of S1 and S2.) Two outputs (at pins 7 and 8) drive LED1 and LED2, which are used as mode indicators. There is also a muting circuit that keeps the LED indicators from flickering.

A monaural signal is fed to the paralleled left and right inputs (pins 2 and 17 respectively) where it splits to follow two signal paths. In one path, the signal is fed through to the output without modification. In the other, all frequencies between 300 Hz and 2 kHz are delayed for a frequency-dependent amount of time (for example, 500μs at 800 Hz) to produce a pseudo-stereo output. In the spatial mode, the stereo effect is enhanced by applying about 50% of the 180° out-of-phase feedback as crosstalk between the right and left channels. Additional information is available from: **Signetics Corp.,** PO Box 409, Sunnyvale, CA 94086.

8-BIT ANALOG-TO-DIGITAL CONVERTER

Analog Devices recently introduced the AD673, a monolithic 8-bit analog-to-digital converter (ADC). This IC provides a microprocessor interface that performs an 8-bit conversion in 30 μs maximum. In addition, it has three-state output latches that are controlled by the data-enable input pin. The data-enable control allows the microprocessor to accept data whenever it is ready. Also included in this IC is a successive-approximation register, a Zener diode reference, a comparator, and a clock. The three-state output buffers specify 250 μs maximum access time. The IC is pin-programmable for either 0 to + 10 volts or ±5 volts full-scale.

The AD673 is powered by two supply voltages; one, a positive 5-volt supply and the other can be either −12- or −15-volts. Its relative accuracy is guaranteed over the specified temperature range at ±½ LSB maximum. Maximum offset drift is also specified as ±½ LSB; maximum gain drift is ±2 LSB. The above specifications apply for temperatures ranging from 0 to +70 °C for the "J" grade and −55 °C to +125 °C for the "S" grade. The IC is available in 20-pin plastic or hermetically sealed ceramic DIP packages.

NEW TECHNICAL DATA BOOK

A new 1080-page databook, *Linear Integrated Circuits and MOSFETs* (SSD-240B) contains detailed technical information on RCA's broad range of linear ICs and small-signal MOSFETs. Included are many ICs for applications in both industrial and consumer electronics devices.

The first section, a general overall guide to available products, provides a complete index of types, photos of the many package types, product classifications and selection guides, listings of special terms and symbols, and a cross-reference showing RCA replacements for many industry types. This section is followed by technical data on devices grouped into categories including Operational Amplifiers, Voltage Comparators, Data-Conversion Circuits, Arrays, Power Control Circuits, Differential Amplifiers, Special-Function Circuits, Automotive Control Circuits, Audio

Circuits, Radio Circuits, TV/CATV Circuits, and MOSFETs.

The final section covers device types supplied for military, aerospace and high-reliability applications and shows dimensional outlines for all package types. It also includes lists of current application notes on ICs and MOSFETs.

NPN TRANSISTOR ARRAYS

RCA recently introduced two NPN transistor-array IC's that are designed for high-frequency applications up to 1.5 GHz. The two devices are high-quality versions of the CA3046 and CA3127E and operate over a temperature range of −55 to +125 °C. Each array consists of five general-purpose silicon NPN transistors on a common substrate. The CA3246E array has three independent transistors, along with two differentially connected devices housed in a 14-pin plastic DIP package. The CA3227E consists of five independent transistors in a 16-pin package.

Each transistor in the array has a gain-bandwidth product (f_T) in excess of 3 GHz. That permits the transistors to operate effectively up to 1.5 GHz for use in VHF amplifiers and mixers, IF converters and amplifiers, and other similar devices. At 25 °C ambient temperature, each transistor in the array can dissipate up to 85 mW. Maximum power dissipation for the package is 425 mW up to an ambient temperature of 75 °C. Above 75 °C, the package must be derated linearly at 6.67 mW/ °C. The breakdown voltage for each transistor are: collector-to-emitter voltage (V_{CEO}) 8 volts, collector-to-substrate voltage (V_{CIO}) 20 volts, and collector-to-base voltage (V_{CBO}) 12 volts. Each unit can handle 20 mA of collector current.

Chapter 24

Regulator ICs

Mention voltage regulators when talking to a group of knowledgeable electronics buffs and you're sure to be met with "So what," "What's new?" or some other similar remark indicating uninterest. After all, most electronics technicians and experimenters consider the voltage regulator to be a simple 3-terminal device that outputs a specified voltage that's lower than its supply voltage. Your "know-it-all" friends wouldn't be so smug had they read Motorola's data sheet on the TL431 adjustable-shunt regulator (introduced by Texas Instruments and now second-sourced by Motorola).

The TL431 is a precision adjustable-shunt regulator, whose output is programmable from 2.5 to 36 volts, using only two external resistors. The current range of that device is from 1 mA to 100 mA, with a low dynamic-impedance of typically 0.22 ohms. Its internal temperature compensation ensures stable operation with a temperature coefficient of 55 ppm/°C. The TL431 is available in three temperature ranges: −55 to +125 °C, −40 to +85 °C, and 0 to 70 °C. Two case styles are available: an 8-pin plastic or ceramic DIP, or the TO-92 case (see in Fig. 24-1).

Some typical applications for the TL431 are shown in Fig. 24-2. Figure 24-2(a) shows a shunt regulator; 24-2(b) is a high-current shunt regulator; 24-2(c) shows a constant-current source, and 24-2(d) shows a triac crowbar circuit. Figures 24-3 and 24-4 show two of the more unusual applications for this regulator IC. In Fig. 24-3, the TL431 is teamed up with an LM11 (precision low-drift op-amp)

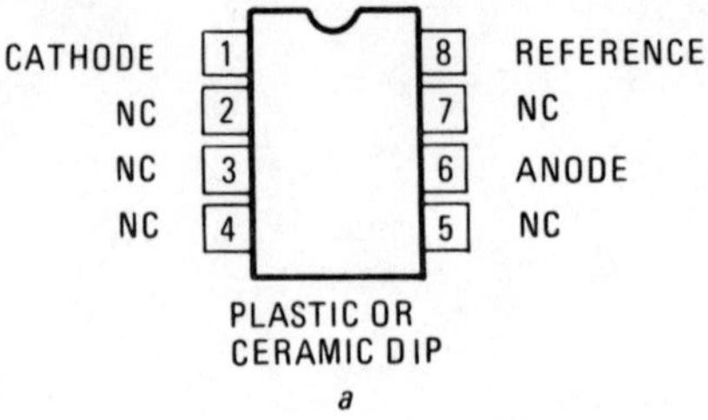

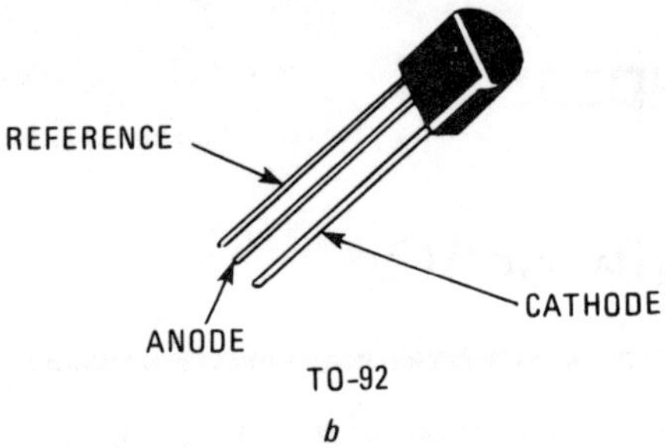

Fig. 24-1.

to form a linear ohmmeter. The schematic in Fig. 24-4 shows an unusual phono amplifier using the TL431. The circuit shown there resembles a vacuum-tube cathode follower or a solid-state emitter follower with the unit's reference electrode corresponding to the grid or base in triode devices. The output of the crystal phono-cartridge (typically 2 volts) modulates the reference voltage. That, in turn, modulates the TL431 cathode-to-anode voltage (V_{CA}) and the audio output is then coupled to the speaker through an output transformer.

DESIGNER'S GUIDE

Designer's Guide—Small-Signal, Low-Noise Transistors is a 126-page data and applications manual covering the TRW line of transistors for receiver and RF circuitry. The manual includes a cross-reference listing nearly 100 competitive transistors and 22 TRW substitutes or equivalents. Also included in the manual are packaged outline drawings with dimensions, complete technical data on the 22 TRW transistors, and application notes including schematics and PC-board patterns. The manual is available from the Semiconductor Division, **TRW Electronics Components Group**, 14520 Aviation Blvd., Lawndale, CA 90260.

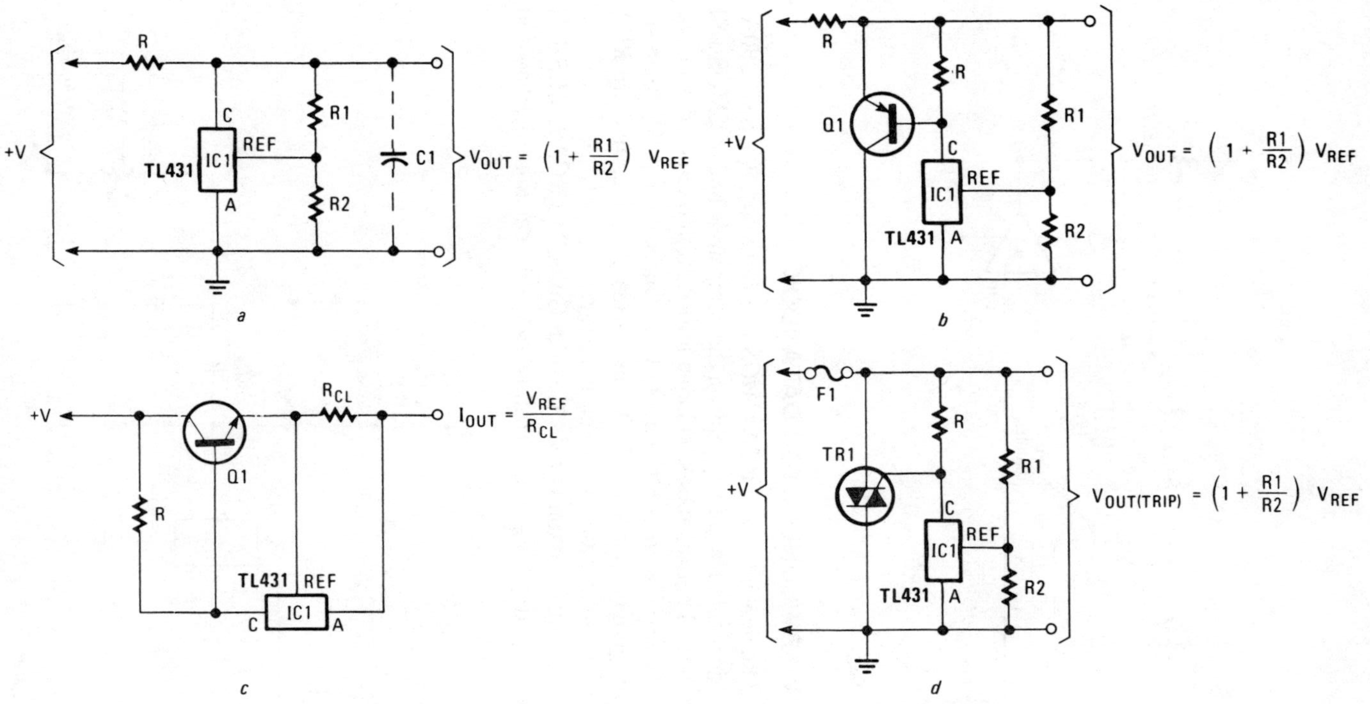
+V
R
C
IC1
TL431
REF
A
R1
R2
C1
VOUT = (1 + R1/R2) VREF
a
R
Q1
C
IC1
REF
TL431
A
R1
R2
+V
VOUT = (1 + R1/R2) VREF
b
+V
Q1
RCL
R
TL431
REF
IC1
C
A
IOUT = VREF/RCL
c
F1
R
TR1
C
IC1
REF
TL431
A
R1
R2
+V
VOUT(TRIP) = (1 + R1/R2) VREF
d

Fig. 24-2.

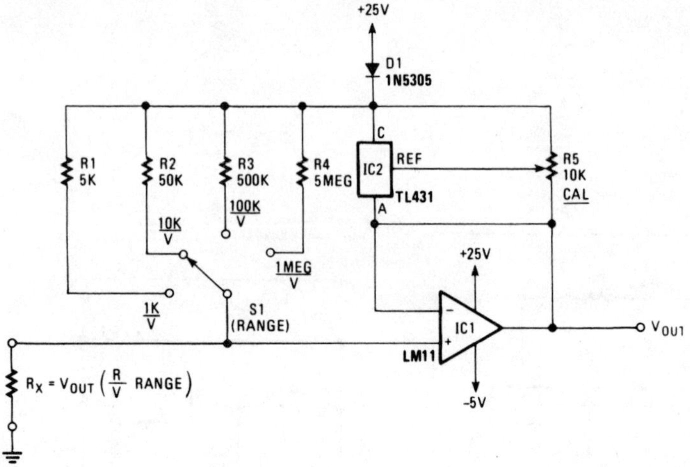

Fig. 24-3.

NEW SEMICONDUCTOR DATABOOK

The new 588-page RCA *CMOS-LSI Databook*, No. SSD-260A, provides technical information on the company's line of CMOS-LSI products. That databook includes information on such devices as the 1800-series and 6805-series of microprocessor products, a series of general-purpose CMOS memories. Also included are RCA alternate-source types for the industry.

The introduction is the "General Guide" to RCA CMOS-LSI products. It points out currently available package options and

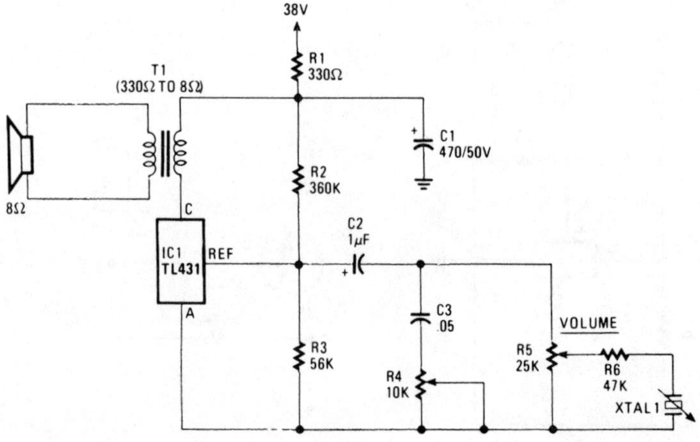

Fig. 24-4.

summarizes the basic features in each product category. In addition, the various ICs are classified according to product type and function.

Five separate data sections provide ratings, electrical characteristics, significant features, and user information for: 1800-series microprocessors and microcomputers, 1800-series memories, 1800-series peripherals, general-purpose memories, and 6505-series LSI products. The *CMOS-LSI Databook* is available from RCA distributors or **RCA** (Solid State Division, Box 3200, Somerville, NJ 08876).

Chapter 25

Power Supply Over/Under Voltage Monitor

Most ICs and transistors have specified "absolute maximum voltages" that, when exceeded, can destroy the device. Likewise, electronic circuits that use such devices have maximum supply-voltage limits. In addition, many solid-state devices also have voltage minimums that, when exceeded, can cause the device to malfunction.

Motorola recently announced the MC3425 and MC3525 monolithic power-supply supervisory circuits, which contain dedicated over- and under-voltage sensing channels. Figure 25-1 shows a block diagram of the MC3425/MC3525. Here we see that the over-voltage and under-voltage detector inputs feed two voltage comparators that are referenced to an internal 2.5-volt bandgap reference.

Separate delay pins (2 and 5) are provided for each channel to independently delay the drive and indicator outputs. That provides greater noise immunity and freedom from false triggering. The two delay pins are really the outputs of their respective input comparators and provide a constant-current source, $I_{DLY(source)}$. The delay pins are pulled low when their respective input comparator's non-inverting input signal level is brought lower than the inverting input. The delay is based on the time it takes for the 200-μA current source to charge an external delay capacitor (C_{DLY}) to the reference voltage.

The delay pins can sink more than 1.8 mA, so the delay

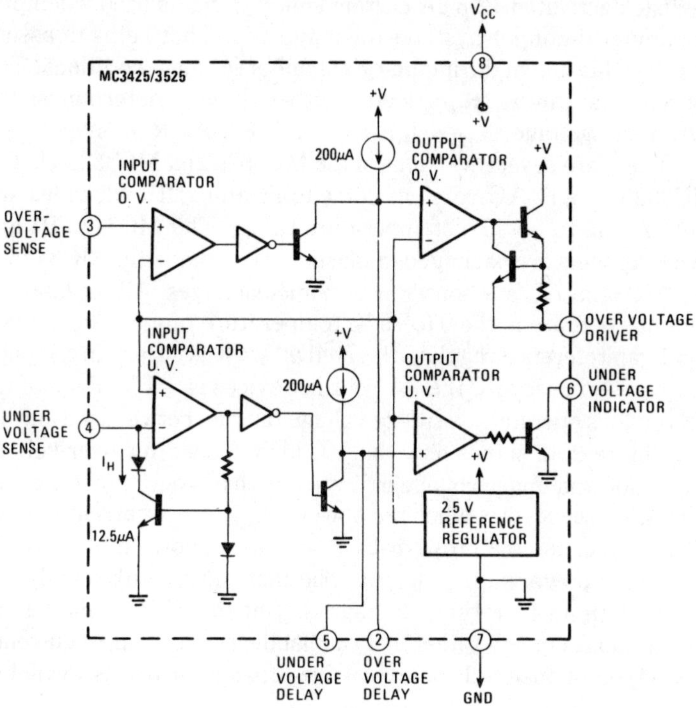

Fig. 25-1.

capacitors can discharge much faster than they can charge from the 200-μA source. When V_{CC} is 15 volts, the delay-time ranges linearly from .001 to 100 milliseconds as C_{DLY} is changed in decade values from .0001 to 10 μF.

The delay pins are also directly coupled to the non-inverting inputs of the output comparators, which are also referenced to 2.5 volts. The output time-delay (t_{DLY}), in milliseconds, is determined by the value of the delay capacitor and is equal to 2.5 $\times$ C_{DLY}/200 or 12,500 C_{DLY}.

The over-voltage channel can source 300-mA with a turn-on slew rate of 2.0 A/microsecond; therefore, it is ideal for driving the gate of an SCR used to "crowbar" the power-supply line leading to a voltage-sensitive device. A current-limited emitter follower drives the over-voltage output at pin 1.

In the under-voltage channel, an open-collector NPN transistor can sink up to 30 mA to drive an LED, small relay, or shut-down circuitry. A fault must be sustained for a predetermined time before there is an output. The under-voltage input comparator has a feedback-activated 12.5 μA current sink (I_H) that is used when programming the input hysteresis voltage, V_H. That helps to assure positive shut-down during marginal undervoltage conditions. The source resistance (R_H), feeding that input, determines the hysteresis voltage: $V_H = I_H R_H = 12.5 \times 10^{-6} R_H$).

There are several versions of the MC3525 and MC3525-U. The MC3525-U and -AU versions come in ceramic DIP packages and have a −55 to +125 °C temperature range. The MC3425-P1 and -AP1 versions are packaged in plastic DIP's, while the MC3425-U and MC3425-AU are housed in ceramic packages. All the last four versions operate in the 0 to 75 °C temperature range. The 2.5-volt band-gap reference regulator is rated at ±4% accuracy in the basic devices. Accuracy for the "A" suffix devices is ±1%. Each of the six versions operate on a supply voltage ranging between 4.5 and 40.

Figure 25-2 shows how the MC3525 is used for over-voltage protection and under-voltage indication with a programmable delay. The fuse is necessary when the power supply is not current-limited. That is because the crowbar SCR acts as a short circuit directly across the power-supply output. The fuse must be extremely fast acting with an I^2t rating less than that of the SCR, while having a continuous current rating that can handle normal supply currents. That type of fuse will be expensive—that is, if one is available.

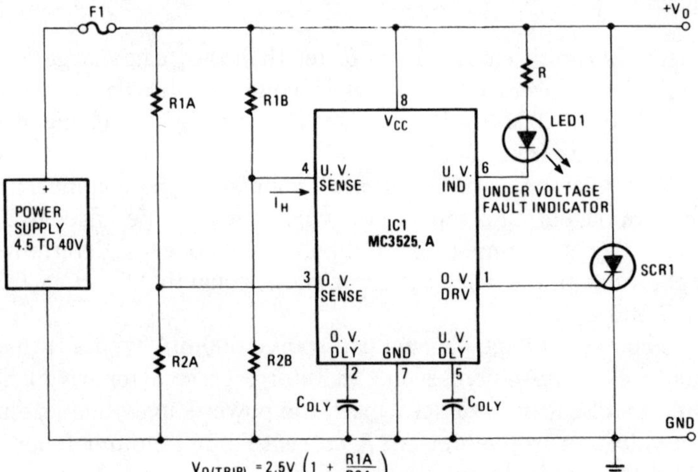

$$V_{O(TRIP)} = 2.5V\left(1 + \frac{R1A}{R2A}\right)$$

$$\text{U.V. HYSTERESIS} = I_H\left(\frac{R1B\ R2B}{R1B + R2B}\right)$$

$$t_{DLY} = 12500\ C_{DLY}$$

Fig. 25-2.

Therefore, we suggest that the protected supply include current limiting in its design. An advance-information data sheet is available from **Motorola Semiconductor Products,** PO Box 20912, Phoenix, AZ 85036.

BI-FET OP-AMP

Analog Devices has introduced a low-cost Bi-FET op-amp (the AD611KH) that offers premium-grade specifications for low prices. The AD611KH delivers a maximum offset voltage of 0.5 mV with an offset-voltage drift of 10 μV/°C, and a guaranteed maximum input bias current of 50 pA.

The lower cost AD611JH version guarantees the maximum offset voltage to be 2.0 mV, with an offset-voltage drift of 20 μV/°C, and an input bias current of 100 pA. The exceptionally low bias current makes the AD611 useful in current-to-voltage conversion applications.

Dynamic specifications for the AD611 include a minimum slew rate of 8 volts/microsecond, a unity-gain bandwidth of 2 MHz, and settling to 0.01% in 3 microseconds. Other specifications include a noise voltage of 2 μV peak-to-peak at 0.1 to 10 Hz, and 0.0025% harmonic distortion. The K versions have a minimum open-loop gain of 94 dB and the J grade has a gain of 90 dB. The devices operate on supply voltages ranging from ±5 to ±18 with a maximum of 2.5 mA quiescent current. They are specified over the 0 °C to +70 °C temperature range, and are housed in 8-pin TO-99 metal packages.—**Analog Devices,** 804 Woburn St., Wilmington, MA 01887.

Chapter 26

A Different Kind of FET

Semiconductor technology has given us some interesting and versatile components. In fact, since their introduction, semiconductors have been continuously upgraded and redesigned so that each new device can do the work of several of its predecessors.

In keeping with that trend, RCA has introduced a new "breed" of MOSFET devices that combines the characteristics of a MOS power transistor, a bipolar transistor, and a thyristor. Called a COMFET—for *C*onductivity-*M*odulated *F*ield *E*ffect *T*ransistor—this device has an on-resistance, $R_{DS(ON)}$, that is substantially lower than that of MOSFET's with the same chip area. That is illustrated in Fig. 26-1.

The COMFET's on-resistance is typically less than 0.1 ohm with a maximum drain current of 20 amperes flowing through the device.

The COMFET, with a chip area of 120 mils^2, blocks 400 volts in the forward direction and 100 volts in the reverse direction. Those characteristics make it suitable for high-voltage, high-power applications. Conventional power MOSFETs are less useful because their high $R_{DS(ON)}$ values cause excessive voltage drops and power losses.

The COMFET acts like a power MOSFET feeding a direct-coupled bipolar compound-connected transistor (see Fig. 26-2). The specifications for COMFET's device are similar to those of bipolar

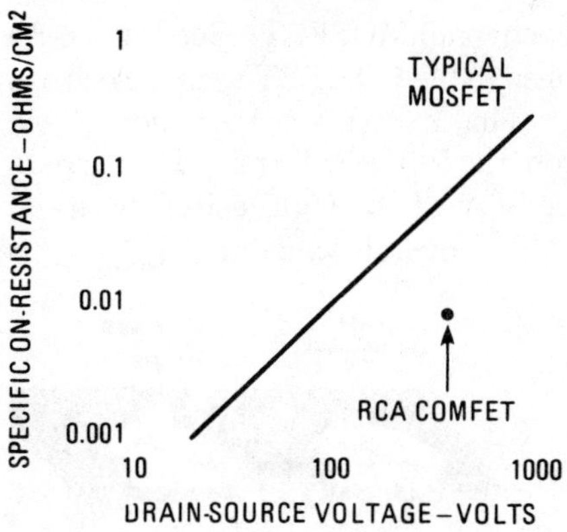

Fig. 26-1.

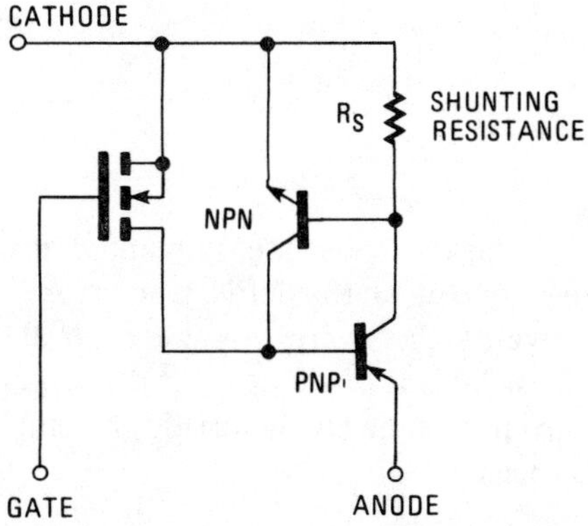

Fig. 26-2.

power-transistors, except that COMFETs have a high input-inpedance, like that usually associated with conventional power MOSFETs. Its high input-impedance makes it simple and easy to drive from a relatively low-power, low-voltage source (such as a logic IC). In addition, they do not require the expensive high-power drivers and complex drive-circuitry that bipolar transistors need. Thus, circuits using COMFETs can be less expensive to operate.

The COMFET's switching speeds are about the same as bipolar devices, but somewhat slower than conventional MOSFET devices. Its typical turn-on time is less than 100 nanoseconds, and turn-off is somewhere between 5-20 microseconds. The COMFET's switching speed and low source-to-drain ON resistance—when considered together—put this device on the same plane as bipolar transistors for most medium-frequency applications.

The device's unusual high-voltage and low-resistance characteristics are produced by applying a P-type substrate on the drain side of a conventional N-channel MOSFET. (See Fig. 26-3.) When a positive voltage is applied to the COMFET's gate, electrons enter the N-type drain region. That causes a corresponding hole to be injected into the drain from the P-type substrate. The carriers or holes modulate the conductivity of the high-resistance drain, thereby substantially reducing the overall value of $R_{DS(ON)}$.

Fig. 26-3.

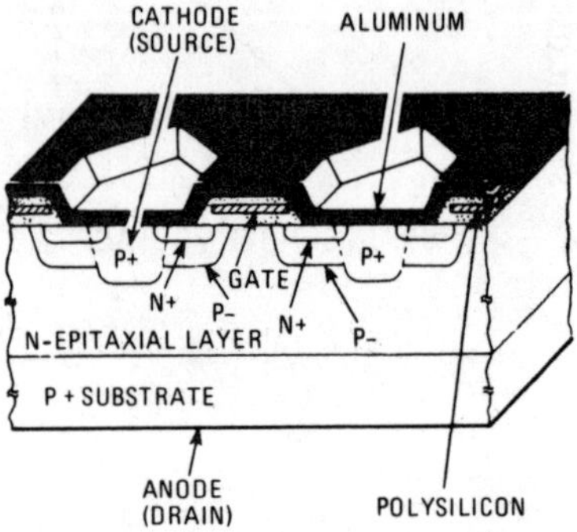

The P+ substrate, implanted at the device's cathode, allows the control of the NPN transistor's shunt resistance (R_S), and prevents the sharp voltage drop at high-current levels that is characteristic of a four-layer thyristor. Therefore, the COMFET's current/voltage characteristics resemble that of an ordinary bipolar transistor.

IF SYSTEM IC

National Semiconductor has announced its latest series of FM IF systems, the LM1865/1965/2065, for use in electronically and manually tuned radio applications.

The LM1865 and LM2065 feature a *stop* detector for electronic tuning. All versions offer a low total-harmonic-distortion (THD) of 0.1% at 100% modulation with a single tuned quadrature coil. And THD is not adversely affected if the radio or quadrature coil is mistuned.

The only difference between the LM1865 and LM2065 is the direction of the AGC control voltage. The LM2065 develops forward AGC. The three devices have dual-threshold AGCs to eliminate the need for a local/distance switch. The AGC threshold-voltage is low when strong out-of-band signals are present—signals that might develop an interfering third-order intermodulation (IM3) product. If there are no strong out-of-band signals present, the AGC threshold is high for best signal-to-noise performance.

Chapter 27

Low-Power Amplifier ICs

If you've been frustrated in your efforts to complete an electronics project or experiment by the lack of a small, low-power audio amplifier, I urge you to consider building one around either of two power-amplifier ICs recently introduced by the Sprague Electronic Company.

First we'll look at the ULN-3705M—its pinout is shown in Fig. 27-1. That class-AB amplifier is housed in an 8-pin mini-DIP and has a voltage gain of 42 dB. Its operating temperature can vary between −20 and +85 °C.

The typical power-output of the unit can range from 60 to600 milliwatts, depending on supply voltage and speaker impedance. For example, operating from a 9-volt supply and driving a 16-ohm load, the unit's output power is 600 milliwatts. Audio power output levels for three different speaker impedances (8, 16, and 32 ohms) and supply voltages appear in Table 27-1.

The recommended supply voltage for the ULN-3705M can range from 4.5 to 9 volts with a 6- to 10-mA quiescent current-drain. However, the amplifier operates (at reduced volume) with supplies as low as 1.8 volts without a notable increase in harmonic distortion. Total harmonic distortion is specified at from 0.4-1.0% and 0.5-1.0% for 32-and 16-ohm loads, respectively.

The ULN-3705M is recommended for use as a headphone driver in battery-powered portable radios and tape recorders. It is, therefore, ideally suited for use in audio signal-tracers and bench

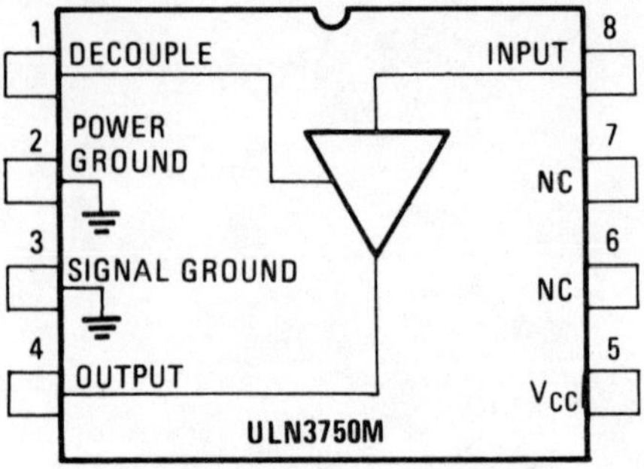

Fig. 27-1.

Table 27-1.

Characteristic	Symbol	Test Conditions	Limits			Units
			Min.	Typ.	Max.	
Supply Voltage Range	V_{CC}		1.8	6.0	9.0	V
Quiescent Supply Current	I_{CC}	$V_{CC} = 4.5$ V	—	6.0	—	mA
		$V_{CC} = 6.0$ V	—	7.0	15	mA
		$V_{CC} = 9.0$ V	—	10	20	mA
Voltage Gain	A_V		—	42	—	dB
Audio Power Output	P_{OUT}	$R_L = 8\Omega$, $V_{CC} = 4.5$ V, THD = 10%	—	220	—	mW
		$R_L = 8\Omega$, $V_{CC} = 6.0$ V, THD = 10%	250	430	—	mW
		$R_L = 16\Omega$, $V_{CC} = 4.5$ V, THD = 10%	—	125	—	mW
		$R_L = 16\Omega$, $V_{CC} = 6.0$ V, THD = 10%	150	240	—	mW
		$R_L = 16\Omega$, $V_{CC} = 9.0$ V, THD = 10%	—	600	—	mW
		$R_L = 32\Omega$, $V_{CC} = 4.5$ V, THD = 10%	—	60	—	mW
		$R_L = 32\Omega$, $V_{CC} = 6.0$ V, THD = 10%	85	110	—	mW
		$R_L = 32\Omega$, $V_{CC} = 9.0$ V, THD = 10%	—	310	—	mW
Distortion	THD	$P_{OUT} = 50$ mW, $R_L = 32\Omega$	—	0.4	1.0	%
		$P_{OUT} = 50$ mW, $R_L = 16\Omega$	—	0.5	—	%
Output Noise	V_{OUT}	Input Shorted, BW = 80kHz	—	225	—	μV
Input Resistance	R_{IN}	Pin 8	—	250 K	—	Ω
Power Supply Rejection	PSRR	C_D (Pin 1) = 500 μF, $f = 120$ Hz	—	34	—	dB

amplifiers, and is intended as a low-cost alternative to designs using discrete transistors.

Figure 27-2 shows a typical application for the ULN3705. The circuit's performance is influenced by the values of two external electrolytic capacitors (C3 and C4). One is used for output coupling and the other for feedback and ripple decoupling.

The output-coupling capacitor, C4, works with the speaker impedance to control the low-frequency cutoff. The −10-dB points are about 20, 30, and 60 Hz when driving an 8-ohm speaker through coupling capacitors of 500 μF, 250 μF, and 100 μF, respectively.

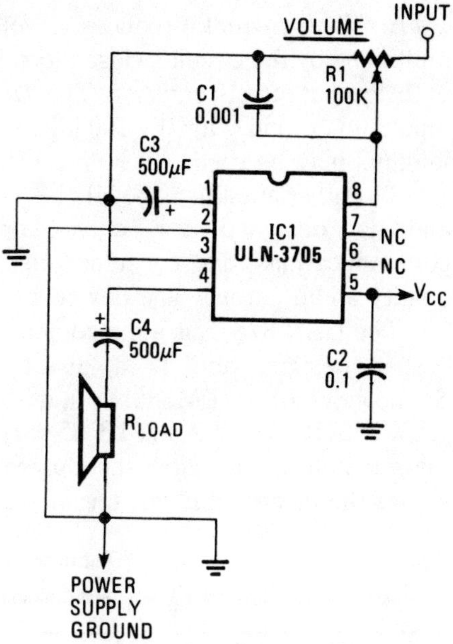

Fig. 27-2.

Capacitor C3 is used for feed-back decoupling and power-supply ripple rejection. The 500-μF capacitor specified provides 34 dB of rejection at 120 Hz. The amplifier input (pin 8) must be provided with a DC return path to ground for a current of approximately

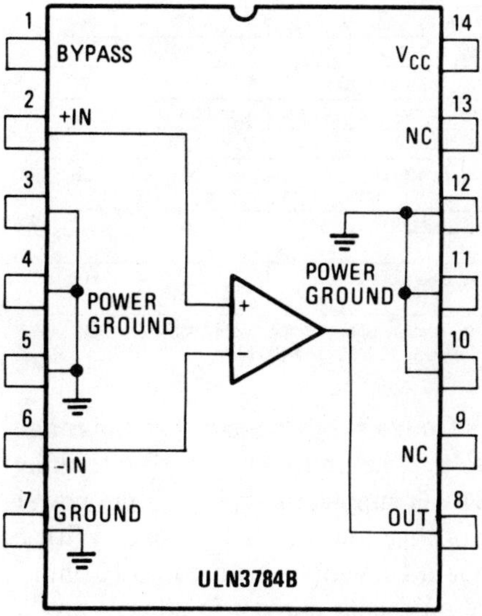

Fig. 27-3.

1 μA. That current produces a voltage (IR) drop that, when multiplied by the circuit's closed-loop DC gain, appears as an error in the output offset-centering. The value of resistance between the input and ground is ideally 100 kilohms or less, although up to 200 kilohms may be used.

The other amplifier, the ULN-3784B, is housed in a 14-pin DIP and is designed for such applications as automotive, communications, and consumer electronics that require a high-quality audio output. The device's pinout is shown in Fig. 27-3.

The ULN-3784B is a pin-compatible improvement on several earlier designs and is a direct replacement for National Semiconductor's LM380N and LM384N, and Sprague's ULN-2280B and ULN-2281B IC's. However, it provides a wider margin of protection against supply-voltage transients. Table 27-2 shows the electrical characteristics of the IC.

Table 27-2.

Characteristic	Symbol	Test Conditions	Limits Min.	Typ.	Max.	Units
Supply Voltage Range	V_{CC}		9.0	24	28	V
Quiescent Supply Current	I_{CC}	$V_{IN} = 0$ V	—	20	—	mA
Quiescent Output Voltage	V_{OQ}	$V_{IN} = 0$ V, See Note 1	—	12	—	V
Voltage Gain	A_V	$P_{OUT} = 0$ W	31	34	37	dB
Total Harmonic Distortion	THD	$P_{OU}T = 50$ mW, $R_L = 8\Omega$, $V_{CC} = 24V$	—	0.2	—	%
		$P_{OU}T = 50$ mW, $R_L = 16\Omega$, $V_{CC} = 28V$	—	<0.2	—	%
		$P_{OU}T = 4$ W, $R_L = 8\Omega$, $V_{CC} = 24V$	—	<0.3	5.0	%
Audio Output Power	P_{OUT}	$R_L = 8\Omega$, $V_{CC} = 24$ V, THD = 5%	4.0	5.0	—	W
		$R_L = 16\Omega$, $V_{CC} = 28$ V, THD = 5%	4.0	4.8	—	W
Input Impedance	Z_{IN}	Each Input	140 K	170 K	—	Ω
Power Supply Rejection	PSRR	$P_{OUT} = 0$ W, $f = 120$ Hz	—	30	—	dB
Equiv. Input Noise Voltage		$f = 20$ Hz to 20kHz	—	60	—	μV_{RMS}
Bandwidth (−3 dB)	BW	$P_{OUT} = 1$ W, See Note 2	—	100	—	kHz

NOTES: 1. The quiescent output voltage typically equals ½ the supply voltage
2. Unity gain typically occurs between 10 MHz and 100 MHz.

The ULN-3784B operates from a single supply that can range from 9 volts up to about 28 volts. The unit has a 34-dB internally fixed gain. Operating from a 24-volt supply, its typical output power is 4-watts minimum when working into an 8-ohm load. With a 28-volt supply, its power output is 4.8 watts working into 16 ohms.

The ULN-3784B's typical quiescent current is 20 mA, and it

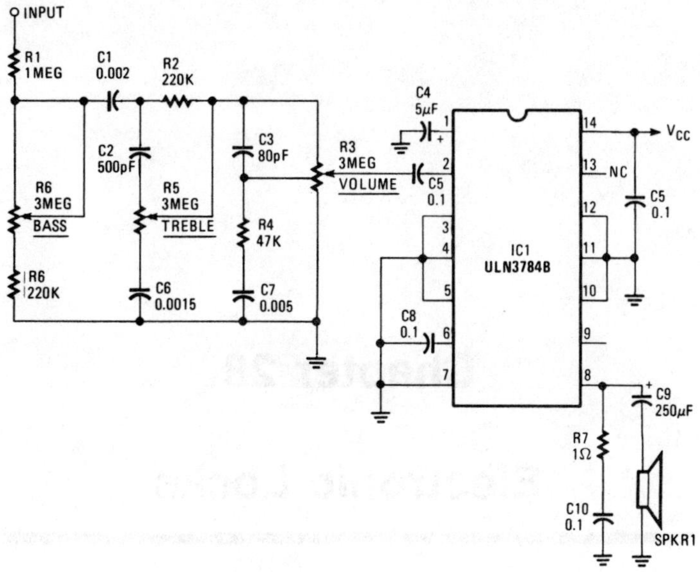

Fig. 27-4.

has a bandwidth of 100 kHz and its power-supply rejection ratio (PSRR) is 30 dB.

Figure 27-4 shows how the ULN-3784B is used as an amplifier with bass and treble controls.

Chapter 28

Electronic Locks

The rapidly rising crime rate has forced us to go to the extremes to protect our valuables. For instance, it is not uncommon to see two and three locks on a single door.

If you use conventional locks to secure your home or valuables, you could find yourself carrying large numbers of keys. That means everytime you want to open a door or gain access to your property, you'll be saddled with the task of sorting through several keys to find the one you need at that particular moment. However, there is another way to get the needed security—replace some of those conventional locks with coded electronic types.

Two electronic locks worth your consideration are the LS7228 and LS7229 from LSI. Figure 28-1 is a block diagram of the innards of the LS7229 (the LS7228 is similar). Both units are ion-implanted, PMOS encoder circuits that include all the necessary logic to interpret the entry code and develop a momentary lock-control output. The LS2778 address decoder is keyed by two pulse trains of logic one's and zero's applied to the correct terminals. The LS7229 is keyed by two double-throw momentary pushbutton switches (which are used to enter one's and zero's).

Both units (housed in 16-pin DIP's) feature stand-alone lock logic, out-of-sequence disabling circuits, current-source lock-control outputs, externally controlled delay to set maximum time between pulses, and a 9-bit entry code determined by 9 parallel inputs. Each

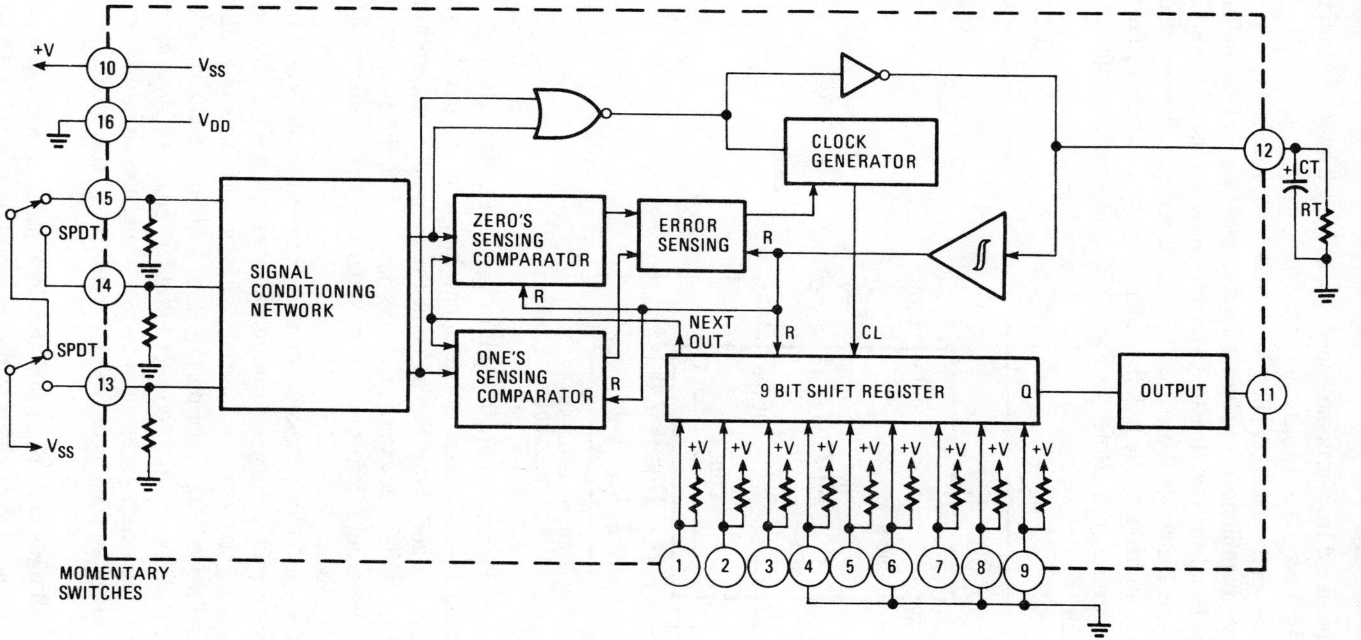
+V
10
VSS
16
VDD
15
14
13
SPDT
SPDT
VSS
MOMENTARY SWITCHES
SIGNAL CONDITIONING NETWORK
ZERO'S SENSING COMPARATOR
ONE'S SENSING COMPARATOR
ERROR SENSING
CLOCK GENERATOR
R
NEXT OUT
CL
9 BIT SHIFT REGISTER
Q
OUTPUT
11
12
CT
RT
+V
1
2
3
4
5
6
7
8
9

Fig. 28-1.

IC is powered from a single-ended 2.5- to 15-volt supply. Maximum standby current is 15 μA.

The locks are controlled by a 9-bit binary code that has 512 possible combinations. The leading or most significant bit is set by pin 1 and the end (least significant) bit is set using pin 9. Code terminals 1 through 9 control a 9-bit shift register. The entry code is programmed into the lock by either jumpering or floating (leaving open) certain pins. Refer to Fig. 28-2, a practical circuit for the LS7229.

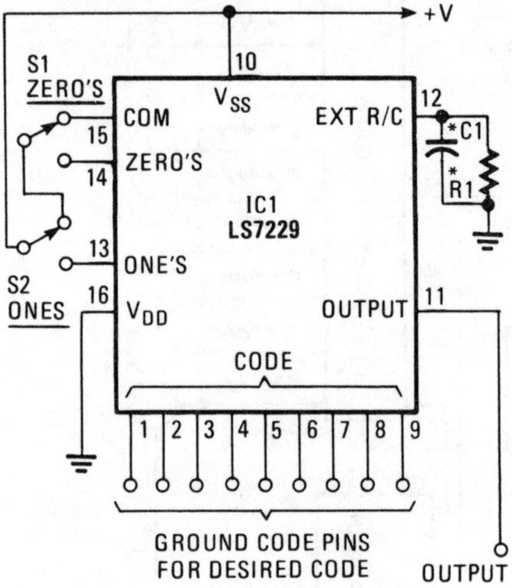

Fig. 28-2.

To program any given input to accept a logic 1, the pin corresponding to that input is left open. Jumpering a pin to ground programs a zero into the device at the corresponding position in the entry code. For example, if pins 3, 4, 5, 6, and 9 are grounded, and the others left open, the binary access-code would be 110000110.

The device is unlocked by entering the code (one's and zero's) in the correct sequence through switches S1 and S2. The zeros' and ones' entry ports are initially at logic zero (ground). As each key is pressed, its entry port goes to logic one and then returns to zero. When the first correct bit of the code is keyed in via S1 and S2, the external capacitor is discharged and an internal inhibit is removed so the circuit will be receptive to the second bit, and so on.

If all nine bits are in the correct sequence,a logic one passes through the shift register to the lock output at pin 11. An out-of-sequence entry or incorrect bit at any point in the entry code inhibits any further entry. After a delay period (determined by the time constant of an external R/C network) a new sequence of key pulses may be applied.

The lock-output (pin 11) switches from zero to logic one as the voltage on pin 9 returns to zero following the last pulse of the entry code. It remains at logic one for a period about 30% longer than the R/C time constant. To hold the output at logic one, apply a tenth entry bit to either pin 13 or pin 14 and hold it high for as long as is necessary.

The output control is a current source so a load must be connected between pin 11 and ground. The source-current range depends on the supply voltage and the voltage across the load. For example, the source current averages 9 mA with a 9-volt supply and 8.5 volts across the load. It sources 26 mA with 7.5 volts across the load.

The time constant of the external R-C network at pin 12 determines the duration of the output pulse and the maximum permissible interval between valid entry-code bits. The time constant in seconds is the product of the resistance of R1 in megohms and the capacitance of C1 in microfarads. When using a 9-volt supply, the minimum suggested value for R1 is 2200 ohms and the maximum value is 3.3 megohms.

Chapter 29

High Power FETs

Until a few years ago, FETs could handle only about 1 watt, and therefore could not compete with SCRs or bipolar power devices. That's because, as shown in Fig. 29-1, the planar technology used has the drain, gate, and source on the top surface of the die.

High-power devices require relatively large metalized terminals to handle the high drain and source currents. Therefore, for increased power-handling capacity, the die size must be increased. Unfortunately, increasing the size results in high inter-electrode capacitances that severely limit the high-frequency response of the device. Further, the high current traveling through the narrow lateral channel develops a significant voltage drop (producing a correspondingly high ON resistance).

However, about seven or eight years ago, a new technology aimed at increasing the current density of the MOSFET was developed. In that structure (dubbed VMOS), the gate and source are side-by-side on the top surface of the die, while the drain is on the bottom.

Impurities are added (by a process called *doping*) to an epitaxial (single crystal) layer of silicon to promote vertical current flow. In addition, some versions of the VMOS device have a V-shaped notch etched through the source and channel regions to enhance vertical current flow.

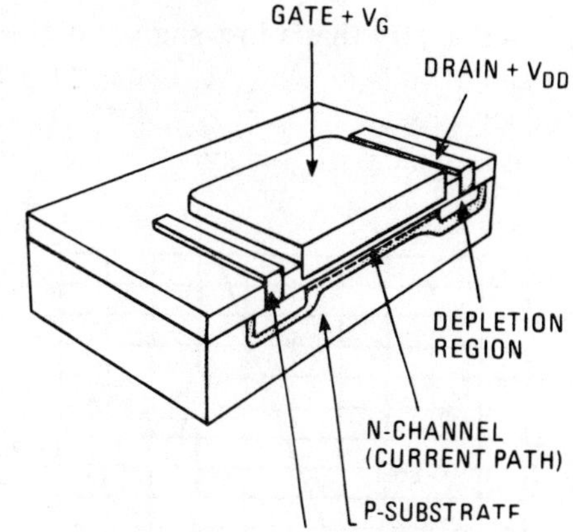

Fig. 29-1.

ENTER TMOS

The TMOS name identifies the technology of Motorola's vertical-channel MOSFETs. Its structure is illustrated in Fig. 29-2. The unique TMOS design, with thousands of "source" sites connected in parallel on a single die, makes possible a very short vertical channel that carries heavy currents, but still has a low ON-state voltage drop. A polysilicon gate material enhances switching speed, while an oxide layer electrically isolates the gate from the source.

The input resistance of TMOS devices is extremely high—10^9 ohms—resulting in a leakage current of only a few nanoamperes. Source-to-drain current is controlled by applying a positive volt-

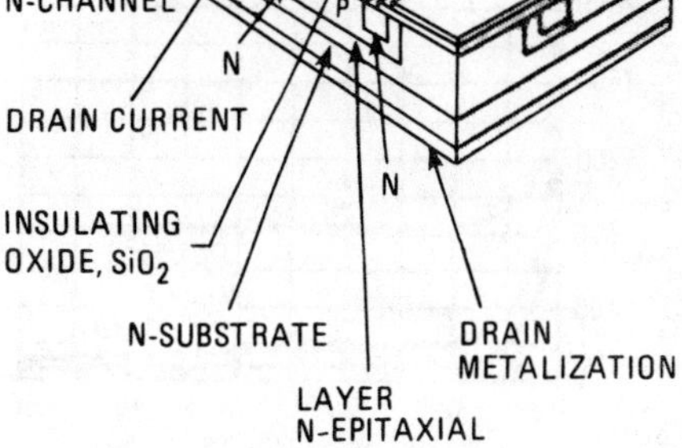

Fig. 29-2.

age to the gate that's just slightly higher than the gate-to-source threshold voltage, $V_{GS(th)}$. The graph in Fig. 29-3 shows the transfer characteristics for Motorola's MTM5N35/40 devices, which is typical of N-channel, enhancement-mode FETs.

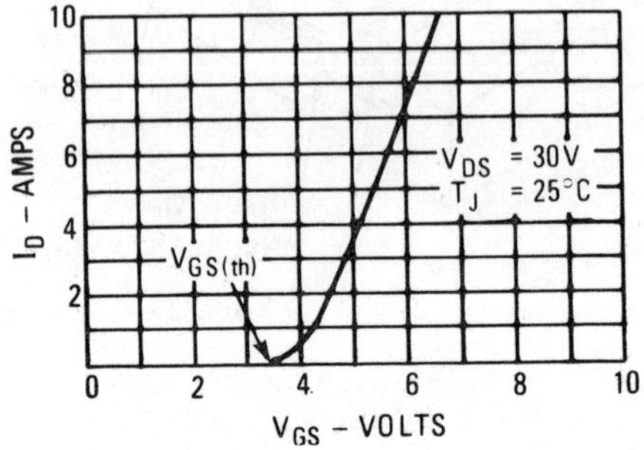

Fig. 29-3.

The extremely high input-impedance and the relatively low gate-drive voltage requirement make the TMOS ideally suited for control of high-power directly from low-level CMOS and TTL logic circuits, if enough current is supplied to charge the input capacitance, C_{iss}.

The TMOS FET has inter-electrode capacitances—gate-to-drain (C_{gd}), drain-to-source (C_{ds}), and gate-to-source (C_{gs})—similar to those of a triode vacuum tube. The input capacitance is $C_{iss} = C_{gd} + C_{gs}$; output capacitance, $C_{oss} = C_{gd} + C_{ds}$; and the reverse transfer capacitance, $C_{rss} = C_{gd}$. Figure 29-4 shows the relationship between those capacitances, which are a function of the drain voltage.

The input capacitance of a MOSFET is important and must be considered in circuit design. On a transient basis, gate current flows to charge C_{iss} before the gate gains control over drain

Fig. 29-4.

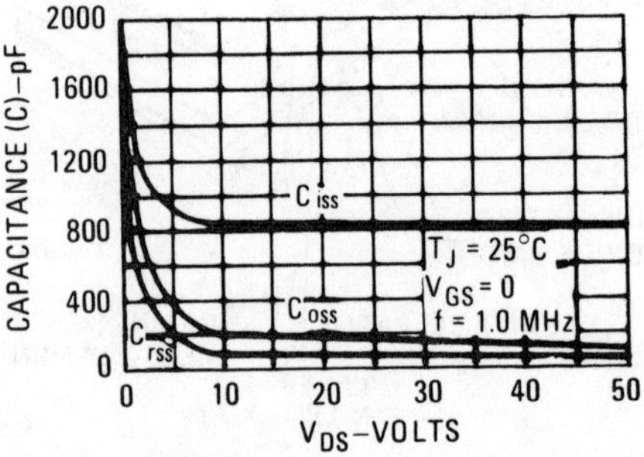

current. The driving (or generator) impedance, R_{gen}, and C_{iss} affect the switching speed. The lower R_{gen} the faster the switching speed.

RCA SOLID-STATE PRODUCT GUIDE

The *Solid-State Devices Product Guide* is a new 56-page publication covering basic descriptions of RCA's extensive product line of solid-state devices. It includes power MOSFET transistors and ultra-fast recovery rectifiers (two new additions to RCA's family of semiconductors).

Other selections include bipolar power-transistors, thyristors, IR emitters, injection lasers, linear and digital logic IC's microprocessors, and numerous other devices. The guide summarizes the devices in each product category and highlights their pertinent features and applications.

LOW BIAS-CURRENT OP AMPS

The 1346 and 1347 are two new high-performance op-amp IC's that combine JFET and bipolar technologies with dielectric isolation to provide superior AC and DC characteristics. Their 2-MHz, unity-gain bandwidth and 7 V/μs slew rate make them ideal for such applications as high-impedance, high-performance buffers, precision track/hold amplifiers, and long-term precision integrators.

Available in TO-00 packages, the 1346 and 1347 are specified over a temperature range of 0 °C to + 75 °C, while the 1346-01 and 1347-01 are specified for the − 55 °C to + 125 °C temperature range. Each unit features ± 250 mA bias current, ± 0.5 millivolt maximum offset, 2μs settling to ± 0.1%, and 300 millivolt maximum power consumption.

Chapter 30

High-Power Microwave FETs

Most power MOSFETs developed so far have been low-frequency devices that are mainly used in switching and audio applications. Thus, silicon bipolar transistors have handled the work load in the design of VHF, UHF, and microwave equipment.

However, it now appears that recent improvements in MOSFET technology will make possible transistors that can handle appreciable amounts of power in the GHz-range, yet provide several advantages over the silicon bipolar transistor. Among the advantages are: DC bias unaffected by temperature fluctuations, higher input-impedance, and lower output-impedance.

The FET is a voltage-driven device so it is ideal for use in Class-A and -AB amplifiers, and other amplifiers that require linear transfer characteristics.

The new high-frequency power MOSFET is called the *P*ower *Si*licon FET) by the developer. It is the result of work advanced by Microwave Semiconductor after J. Nishizawa of Mitsubishi proposed that a JFET with a very short vertical channel would provide superior microwave performance.

The PSIFET uses a short-channel vertical JFET configuration similar to that of the TMOS FET described in the February column. Like other vertical-channel FETs, both gate and source are formed on the top surface of the silicon die and the drain is on the bottom as shown in Fig. 30-1.

In the conventional JFET, the gate voltage controls the cross-

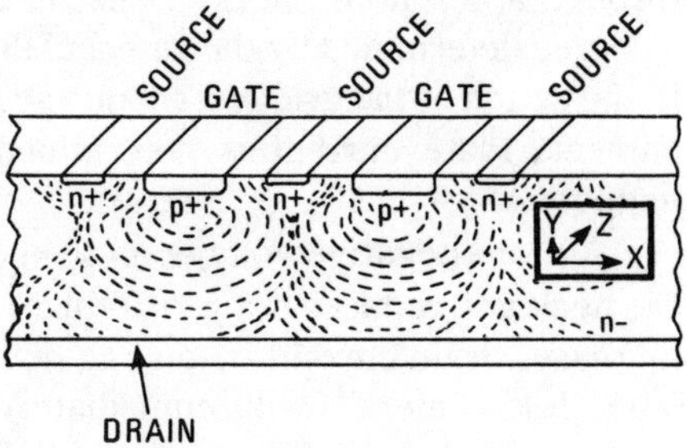

Fig. 30-1.

section of the lateral source-to-drain channel; thus it also controls the channel's resistance and the current flowing through it. Note that in the PSIFET, the channel is always fully depleted (completely devoid of holes and free electrons), although the electric potential in the depletion area varies with gate voltage.

The electric potential in the channel's depletion region forms a barrier that tends to block electron flow between the source and drain. However, if a sufficiently high voltage is applied to the drain, it generates an electrostatic field that counteracts the potential-barrier in the depletion area. That allows electrons to be injected from the source into the drain.

Figure 30-2 shows the distribution of the electric potential in the PSIFET. We can see that the source-to-drain current must flow

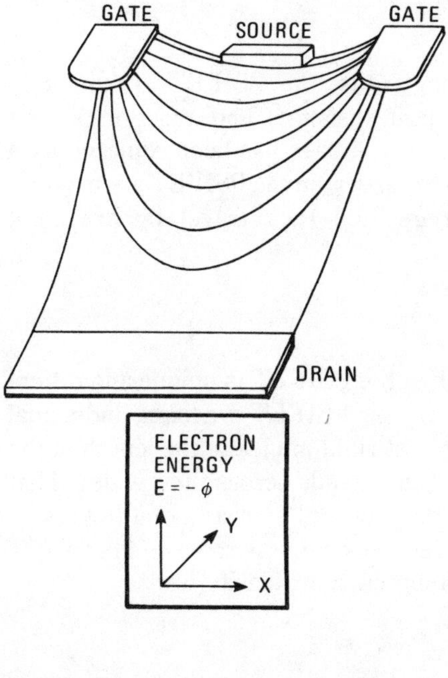

Fig. 30-2.

through a potential "neck." That neck impedes electron flow to a degree determined by the gate and drain voltages. (That's similar to the action of the grid of a triode vacuum tube in controlling plate current. Plate current is determined by the grid and cathode voltages.)

A more negative voltage on the gate increase the potential in the neck and reduces drain current. A more positive drain voltage increases drain current by pulling down the potential in the neck. You "old timers" will immediately recognize the triode-like performance of the voltage-current characteristics curves in Fig. 30-3.

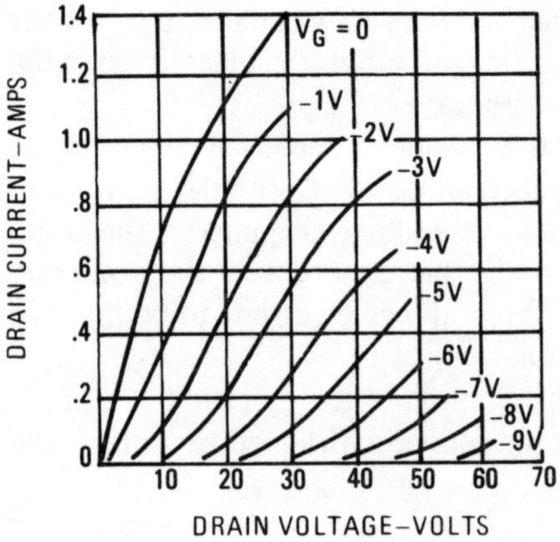

Fig. 30-3.

Despite its non-saturating traits, the PSIFET is similar in operation to a silicon bipolar-transistor. The most important difference is that the neutral p-type base has been replaced by a depleted n-type region. Injected carriers in the PSIFET are majority carriers—not minority carriers—and they travel by drift, not diffusion.

PSIFET DESIGN

The design configuration of the PSIFET is unique and differs from other vertical FETs. In the PSIFET's design, individual source sites (see Fig. 29-4) run at right angles to the length of the source metallization fingers and extends across their width. That layout allows the use of wide gate and source metal-fingers as compared to the 1 to 2 fingers in other microwave silicon fET designs. The wide fingers keep current density low.

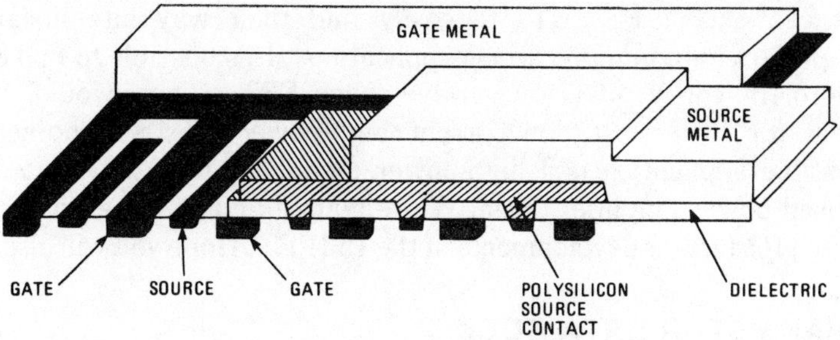

Fig. 30-4.

A highly n-doped polysilicon material provides the intimate contact with the source. That removes a metal contact from the sensitive source area and increases reliability during long periods of high-temperature, high-current operation.

ADVANTAGES OF THE PSIFET

Since its carriers travel by drift rather than diffusion, a PSIFET's carrier velocity is higher than that of a bipolar device and can exhibit a higher f_t (gain-bandwidth product) along with a high breakdown voltage. The PSIFET is relatively insensitive to both long-term and transient exposure to high-energy radiation.

It has a negative temperature co-efficient so it is immune to thermal runaway and hot spots at high current levels. The drain area of the PSIFET is spread out on the back side of the die, where it can be closely coupled to a heatsink. The configuration of the new device does not include the thin SiO_2 insulating film around the gate area, so it is immune to catastrophic failure due to accumulated static charges.

Extensive tests have been run on PSIFET's at 450 MHz and 1 GHz. Input and output microstrip circuits were used to match the devices to 50-ohm impedances. For the 450-MHz test, a PSIFET with a 5.6 cm gate-width and eight separate transistor cells, was formed on a 33 × 75 mil die. The transfer characteristics of the device are in Fig. 30-3. That single die delivers 48.5 watts of saturated CW power with 60% drain efficiency. A similar device with 3.5-cm gate fingers was tested at 1 GHz, and it delivered 21.5 watts saturated CW power at 40% drain efficiency.

APPLICATIONS

Unless it is quickly superseded by vastly improved technology, we can expect PSIFETs to easily find their way into linear amplifiers for communications applications. It is possible to make 70- to 100-volt V_{DD} CW microwave-power FETs. These would be ideal for lower-cost, low-current high-power linears. Another possible application is a high-power modulator for a modulated pulsed-power amplifier chain. We're going to go on the alert and watch for further developments of the PSIFET. Hope you will also.

TRANSISTOR DATA BOOK

The *Small-Signal Transistor Data Book* is an all-new 1350-page databook providing the complete specifications and typical performance curves on the more than 1750 Motorola bipolar and fET devices.

The devices covered are packaged in metal-can, plastic, and miniature housings and are grouped together by families.

The first section is the "Selector Guide," which indexes the transistors into groups according to the headings of the six following chapters. Each chapter has at least three tables listing each device according to a sub-grouping. For example, tables listed in the second chapter ("Plastic-Encapsulated Transistors") are general-purpose, high-speed switching, RF/UHF/VHF amplifiers, high-voltage, dual diodes, and choppers. The five following chapters are: "Microminiature Products" (SOT-23 and SOT-89) which covers those devices contained in surface-mount packaging.

"Metal-Packaged Transistors" covers those devices in TO-18, TO-39, TO-46, TO-52, and TO-72 package options. Included are general-purpose, switching, high-voltage, choppers, Darlingtons, low-noise amplifiers, and RF amplifiers.

"Multiple Transistors" covers both quad and dual transistors that have been implemented on a common substrate.

"Field-Effect Transistors" includes devices operating from DC to UHF for switching and amplifying applications. They are available in plastic and metal-can packages. "RF Transistors" characterizes small-signal high-frequency transistors as low-noise amplifiers, oscillators,high-speed switches, Class A linear amplifiers, and Class C amplifiers. Package types include plastic/ceramic stripline and various metal cans.

The wrap-up is an 8-page section of package outline drawings and dimensions and application information.

Chapter 31

A Light Dimmer

Electronic light dimmers have been used in the home and industry for a couple of decades. Typically, they are relatively simple devices with which the operator controls the level of illumination by using a potentiometer to vary the triggering angle of a triac or SCR that is wired in series with a lamp. The newest development in electronic light dimmers uses a "touch switch" to control the light—both to turn it on and off, and to select the brightness level.

The heart of the new circuit is the LS7237 monolithic PMOS touch-sensitive light dimmer and switch developed by LSI Computer Systems. The block diagram of the device is in Fig. 31-1, and a practical circuit using the device is shown in Fig. 31-2.

The output at pin 8 of the LS7237 is used to trigger a triac. The lamp brightness is, in turn, controlled by varying the output phase angle (triac triggering angle) with respect to the phase and frequency of the AC line.

The output at pin 8 is a low-level pulse of fixed duration (40 to 55 μs) occurring every half-cycle of the AC line voltage. The phase angle (θ) of that pulse, with respect to the AC line, controls the lamp brightness. The phase angle can be varied by applying a low-level pulse to the SENSOR input (pin 5) or a high-level pulse to the SLAVE input (pin 6). That's done by touching the appropriate sensor plate. (When using the SLAVE input, you can use a mechanical push-button or the electronic switch and sensor plate as shown in Fig. 31-3.

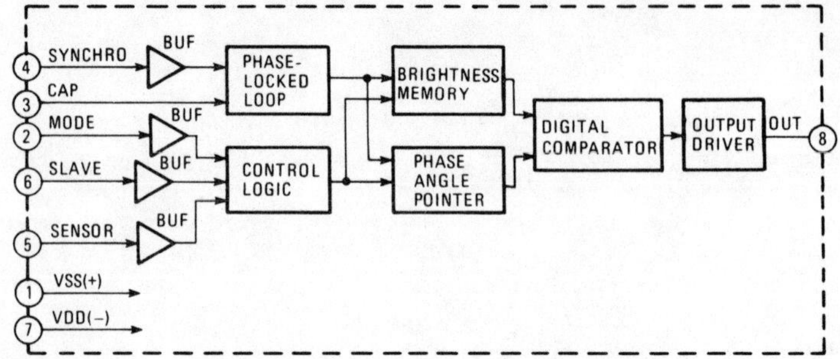

Fig. 31-1.

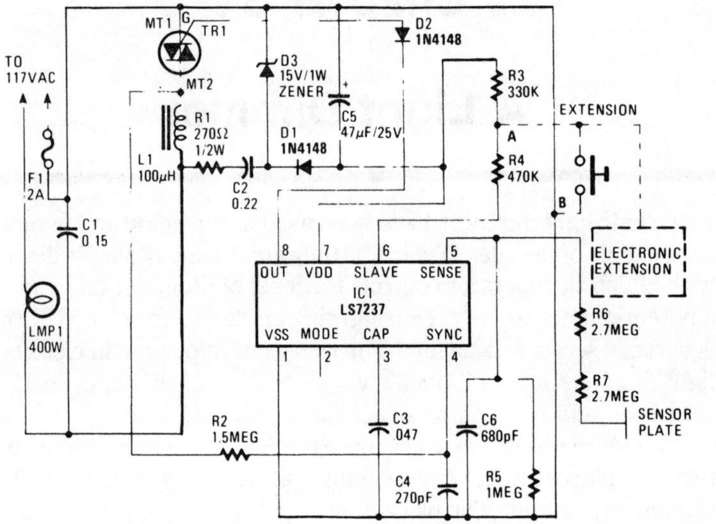

Fig. 31-2.

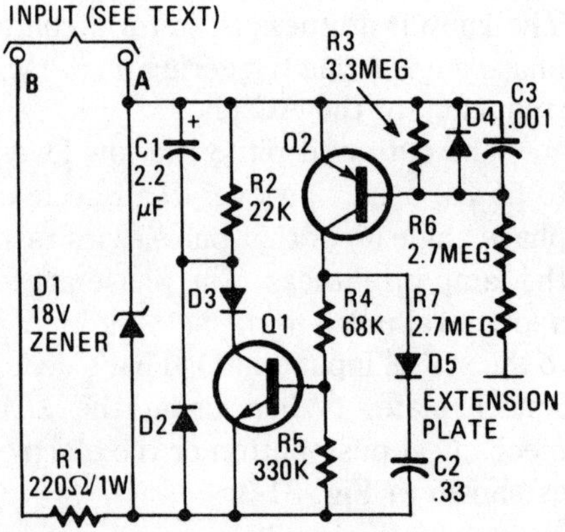

Fig. 31-3.

The relationship between output pulse phase angle, light level, and brightness as a percentage of rated wattage is shown in Table 31-1. The circuit can be set up to operate in one of three different sequences of brightness by tying the MODE pin (pin 2) to a specific voltage level as shown in Table 31-2.

Table 31-1.

Output phase angle (θ)	Brightness levels	Percent rated wattage
No output	Off	0
45°	Night light	9%
70°	Mood light	29%
105°	Medium	66%
159°	Maximum	99%

Table 31-2.

Mode	Mode input	Brightness sequence
0	V_{SS}	Off—Max—Off
1	V_{DD}	Off—Mood—Med—Max—Off
2	Floating	Off—Night Light—Mood—Med—Max

When power is first applied, the output comes up in the OFF state. Then each time the sensor plate is touched, the output steps to the next brightness level.

All input timings for the LS7237 are synchronized with the AC line when this 50/60-Hz source is tied to the SYNCHRO terminal (pin 4) through a suitable lowpass filter network (R2-C4).

Low-level negative-going pulses applied to pin 5 of the LS7237—the SENSE input—cause the circuit to step the lamp through the sequences of brightness as determined by the voltage on the MODE terminal. Output stepping occurs on the trailing edges of the input pulses.

The function of pin 6, the SLAVE input, is similar to the SENSE input except that it operates on positive-going pulses instead of negative ones. It is to be used when long extension cables are used between the sensor plate and the dimmer circuit. In such a case, the electronic extension and extension plate circuit, whose schematic is shown in Fig. 31-3, should be used.

A PRACTICAL CIRCUIT

Let's now return to Fig. 31-2 and see how the components of the basic circuit function.

- Diode D1, R1, C2,C5, and Zener diode D3 deliver a regulated

+15 volts to the V_{ss} terminal of the device.

- Network R2-C4 feeds a filtered 60-Hz signal into the device to synchronize the internal phase-locked loop (PLL) with the line frequency.
- Resistors R3 and R4 limit current in the event that the extension circuit is incorrectly polarized. When the extension touch plate is not used, tie pin 6 directly to ground at pin 7.
- Resistor R5 sets the sensitivity of the sensor input. Select a value between 1 megohm and 5 megohms.
- Capacitor C3 is the filter for the internal PLL.
- Diode D2 limits the positive swing of the triac gate to V_{ss} + 0.5 volt. That positive swing may occur in some triacs when they are triggered.
- Inductor L1 and C1 form an RF filter to suppress RFI developed the triac.

SEMICONDUCTOR DATA BOOK

The *RF and Power Semiconductors Data Book* from TRW contains over 600 pages of electrical specifications, performance parameters, and photographs of nearly 400 semiconductor devices. The semiconductor devices covered in the data book include Zener and Schottky diodes, microwave transistors, and hybrid amplifiers for CATV and VHF-UHF linear applications. A valuable aid in selecting a suitable device for a particular application.

Chapter 32

Automotive Electronics

Electronic ignitions for automobiles have been around for nearly twenty years and are now making their debut in small gasoline engines like those used in lawn mowers, garden tractors, snow blowers, and the like.

In the modern breakerless electronic ignition system, a voltage pulse is induced in a magnetic pickup (reluctor). That pulse is used to gate a power transistor or SCR and thus control current flow through the ignition coil.

Motorola's MC3334 monolithic high-energy ignition system is the latest addition to a line of IC's for automotive applications. It's intended for use in after-market 5-terminal ignition systems.

Figure 32-1 shows a typical application using the MC3334. That circuit takes the control signal developed by the magnetic pickup and processes it to supply a precisely controlled drive to an external Darlington power transistor. When the transistor is turned on, it permits current to build up in the primary of the ignition coil.

Then, at the proper time, the transistor is turned off and current flow is interrupted. The collapse of the magnetic field surrounding the ignition-coil primary produces a flyback action that generates the high secondary voltage needed for the spark.

The MC3334 features an automatically adjusted dwell to produce energy at all engine speeds. It also features a wide supply-voltage operating range (V_{BATT} = 4.0 to 24 volts), over-voltage shut-down at 30 volts (to prevent component damage by transients

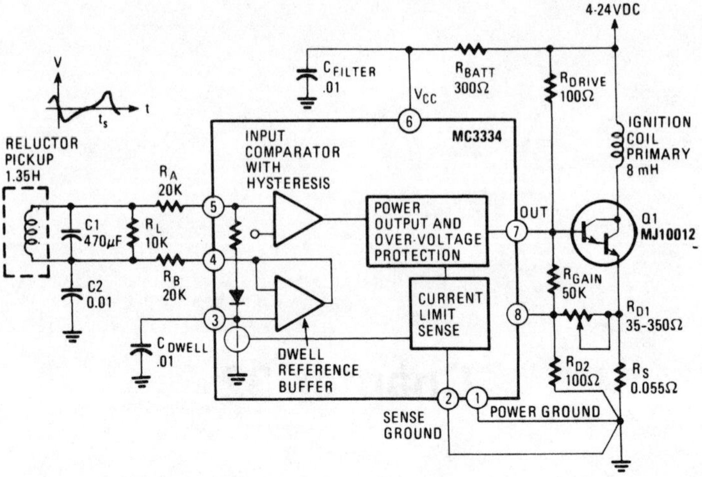

Fig. 32-1.

and excessive battery voltage due to load-dump) and low eternal component count with no critical resistors.

The MC3334 provides a variable DC reference voltage that is used to vary the dwell angle in response to variations in engine speed. (If the dwell angle is fixed, it produces a fixed duty cycle. That results in too little stored energy at high RPM's and wasted power at low engine speeds.) The variable reference voltage is developed and stored across capacitor C_{DWELL}. It's then buffered and fed to the bottom end of the reluctor (pickup) coil.

At high RPMs, the MC3334 holds the output off for around 1 millisecond to permit the energy stored in the spark coil to fully discharge across the spark plug before it turns on the transistor. At that point, ignition coil current begins to increase linearly at a rate determined by the combination of V_{BATT} and the ignition coil's resistance.

As the engine slows, ignition coil-current increases and is limited only by the coil resistance. When the MC3334 is used, peak coil current can be set by an external adjustment, potentiometer R_{D1}. Resistor R_S senses the coil current.

Figure 32-2 shows that when the coil primary resistance is 0.42 ohm, R_{D1} is adjusted to limit coil current to about 5.5 amperes when engine speed drops below about 4000 RPM. For instance, a frequency of 333 Hz corresponds to 5000 RPM for an 8-cylinder, 4-cycle engine.

As engine speed continues to drop, the coil current is held at

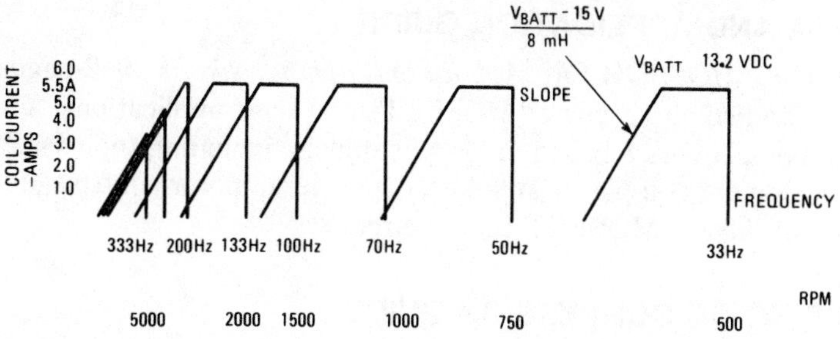

Fig. 32-2.

5.5 amperes for a short period to ensure that there is enough electromagnetic energy in reserve for sudden acceleration.

When battery voltage is low and the engine turns over slowly—during cold weather, for example—the transistor switches on for longer periods to compensate for the slower current build-up. Under those conditions, the ON period may be stretched for nearly a full cycle. Table 32-1 describes the functions of the external components and gives some typical values.

Table 32-1.

PICKUP	—Series resistance = 800Ω ± 10% @ 25°C Inductance = 1.35H@1.0 KHz@ 15 V_{rms}.
IGNITION COIL	—Leakage L = 0.6 mH. Primary R = 0.43 Ω ± 5% @ 25°C Primary L = 7.5 to 8.5 mH @ 5.0A
R_L	—Pick-up load resistor = 10 K ± 20%
R_A, R_B	—Input buffer resistors, provide additional transient protection to the already clamped inputs = 20K ± 20%
C1, C2	—For reduction of high frequency noise and spark transients induced in pick-up and leads, optional and non-critical
R_{BATT}	—Provides load dump protection (but small enough to allow operation at V_{BATT} = 4.0 V) = 300 Ω ± 20%
C_{FILTER}	—Transient filter on V_{CC} non-critical
C_{DWELL}	—Stores reference, circuit designed for 0.1 μF ± 20%
R_{GAIN}	—R_{GAIN}/R_{D1} sets the DC gain of the current regulator = 5.0 K ± 20%
R_{D2}	—R_{D2}/R_{D1} set up voltage feedback from R_s
R_S	—Sense resistor (PdAg in thick film techniques) = 0.055 Ω ± 50%
R_{DRIVE}	—Low enough to supply drive to the output Darlington, high enough to keep $V_{CE(at)}$ of the IC below Darlington turn-on during load dump = 100 Ω ± 20%, 5.0 W
R_{D1}	—Starting with 35 Ω assures less than 5.5 A, increasing as required to set 5.5 A

$$R_{D1} = \frac{I_{o(pk)} R_S - V_{ref}}{\frac{V_{ref}}{R_{D2}} - \frac{1.4}{R_{GAIN}}} - 100\ \Omega(\text{nom})$$

DATA AND APPLICATION GUIDE

The *MOSPOWER Applications Handbook* is 512-pages, dedicated solely to power MOSFETs and their applications. The handbook covers a broad range of subjects, including topics such as design data on power supplies, converters, motor controls, and state-of-the-art MOSFET power circuits.

RMS TO DC CONVERSION GUIDE

The *RMS to DC Conversion Application Guide* is a 56-page softcover booklet that covers RMS-to-DC conversion theory, RMS design considerations, and applications circuits for Analog Devices' AD536, AD636, and AD737 converters. The booklet also covers conversion methods like absolute deviation, and explicit and implicit computation. Also included are discussions on audio processing, modem line monitors, and RMS panel meters.

Chapter 33

A New Series of CMOS Counters

Counters are among the most useful digital-logic circuits. They can receive uniform pulses (representing units to be counted) and develop a voltage that is proportional to the number of pulses input. That makes them useful for such applications as measuring time, period, or frequency.

LSI Computer Systems has introduced a new series of complementary CMOS divider ICs (a divider, of course, is a counter that produces an output for every n^{th} input pulse, instead of for every input pulse). The five new devices, which have been dubbed the RED series, and their functions are: RED $^{5}/_{6}$, divide by 5 or 6; RED 50/60, divide by 50 or 60; RED $^{100}/_{120}$, divide by 100 or 120; RED $^{300}/_{360}$, divide by 300 or 360, and RED $^{3000}/_{3600}$, divide by 3000 or 3600.

The counters are used as timebase generators operating from either 50- or 60-Hz line frequencies. They produce outputs ranging from 10 pulses per second (RED $^{5}/_{6}$) to 1 pulse per minute (RED $^{3000}/_{3600}$), depending on the particular IC.

The devices feature low power dissipation and high noise immunity and can operate from supplies of from +5 to +15 volts. The output is low-power TTL compatible when the IC is operated from a +5-volt supply. The clock input can accept a 50 or 60 Hz sinewave directly thanks to the inclusion of an onboard pulse shaper. The output is in squarewave form, except for the RED $^{5}/_{6}$ when used in the divide by 5 mode. Inputs, in addition to the clock input, include 50-/60-Hz division select, input enable, and reset. All of the inputs are protected.

Figure 33-1 is a pinout of the RED series counters. The devices are identical in all but one respect—the number of Johnson-type decade counters used in the output. For instance the RED $^{5}/_{6}$ uses a single counter, while the RED $^{50}/_{60}$ uses two counters. Figure 33-2 shows the counter-timing diagram. Note that the counting will always be proper if, as at points A through E, the reset and enable signals are allowed to switch only when the input clock is low.

The counter advances by one count on each negative transition of the input clock, as long as the ENABLE is high and RESET is low. When ENABLE is low, the input clock pulses are inhibited and the counter is held in the state it was in prior to switching. A high reset signal clears the counter to zero. A low on the DIVISION SELECTION pin causes a divide by 6, 60, 120, 360, or 3600, depending on the counter being used. A high on the DIVISION SELECT pin causes a divide by 5, 50, 100, 300, or 3000.

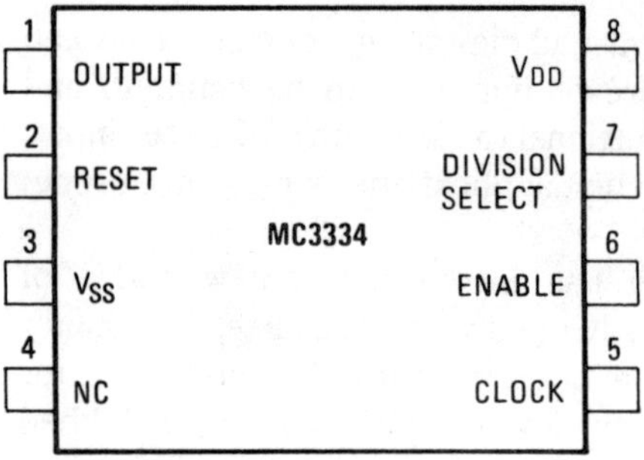

Fig. 33-1.

NEW LOW-DROPOUT REGULATOR

A positive voltage regulator that can source output-current up to 1.5 amperes with an input/output differential of typically 0.65 volt is now available from National Semiconductor. The latest addition to their line of regulators, the LM2940CT-5.0, features an output current in excess of 1 amp that, along with the low drop-out voltage, makes it useful in applications where the input voltage is kept at a level within a volt or two of the output.

The device is a plug-in replacement for similar NPN-type regulators and is pin-for-pin compatible with the LM340. Its guaranteed accuracy is $\pm 3\%$ at 25 °C. The LM2940CT-5.0 is ideal for use in battery-powered applications and in automotive

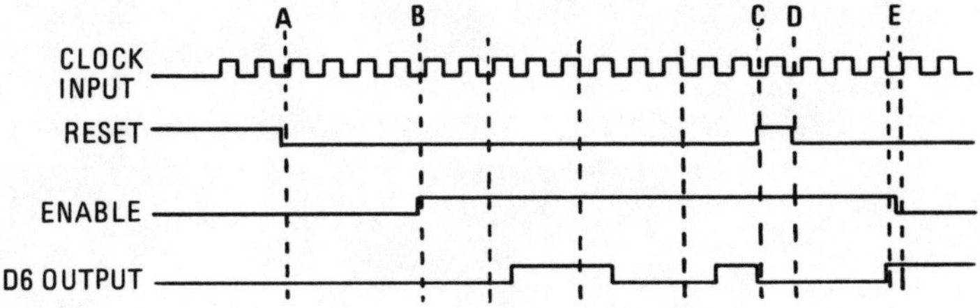

Fig. 33-2.

applications where large transient voltages, such as ±45 volts, can be expected.

Its features include internal short-circuit current limiting, thermal-overload protection, low quiescent current, reverse-battery protection, and a trimmed output voltage. Other voltage options are soon to be made available.

CMOS AND TMOS TECHNOLOGIES COMBINED

The *SMARTpower II* device is a Motorola innovation that combines the best qualities of high-speed CMOS logic and high-current TMOS vertical power structuring into a single IC.

The MPC2005 can handle up to 15 amperes of continuous anode current, but is also capable of discharging capacitors with peak currents to 150 amps. When voltages exceed 6.2 volts or the junction temperature rises above 125 °C, the device trips and removes the voltage from the system.

The MPC2005 can protect voltage-sensitive systems and memories on 5-volt buses from overvoltage transients and extreme temperatures that can damage the system. The device dissipates extremely low amounts of power in the off state, hence, it can be used to protect temperature sensitive devices since its junction temperature is essentially the same as that of the case.

The device includes a pin that can be used to set the trip-point to a level other than the pre-set 6.2 volts. A capacitor can be connected to the pin to change the delay time of the trip voltage. A single MPC2005 can replace three resistors, an SCR, and an IC.

Chapter 34

A New COMFET

The world's first P-channel COMFET has been announced by RCA's Solid-State Division and RCA Labs at Princeton, NJ. The new COMFET (*C*onductivity *M*odulated *F*ield-*E*ffect *T*ransistor) complements the N-channel COMFET. Switching speed, forward blocking voltage, and on-resistance characteristics are similar for the N- and P-channel types.

An unusual feature of the P-channel COMFET is that its chip area is the same as that of the N-channel device. Normally, in conventional MOSFET technology, the P-channel devices require a larger chip area than N-channel devices because of the mobility of "holes" (the primary current carriers) is less than the mobility of electrons (the primary current carriers in a conventional N-channel MOSFET).

In the new device, the conductivity modulation effect nullifies the mobility factor, permitting both chips to have the same on-state resistance with the same size die. Since they have identical active areas, their input capacitances are identical, thus permitting their use in complementary applications found in many types of power-control circuits. The P-channel device can also be used alone in variety of power-switching applications.

Another noteworthy characteristic of the P-channel COMFET is its extremely low DC on-resistance (0.35 ohm at 20 amps) at forward blocking-voltages between 200 and 400 volts (see Fig. 34-1). That is the lowest on-resistance of any P-channel power

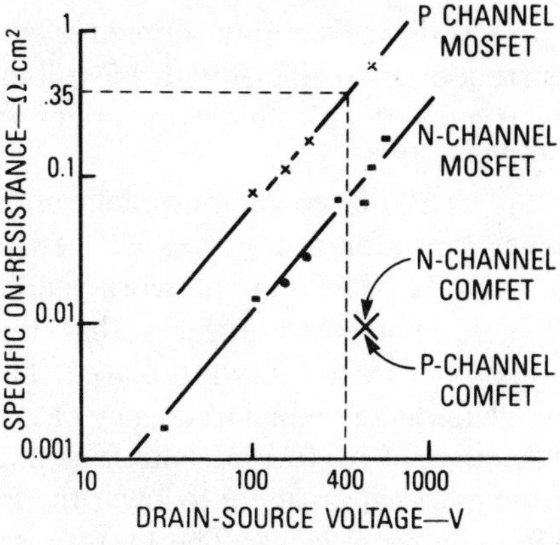

Fig. 34-1.

MOSFET, including those with blocking voltages of only 100 volts. Furthermore, its on-resistance can be as much as 25 times lower than that of a conventional P-channel MOSFET having a comparable blocking voltage.

Generally, a COMFET and a conventional power MOSFET operate in basically the same manner. However, in a conventional MOSFET, the on-resistance increases with the increasing drain-to-source voltage capability. That limits its use as a power-switching device above 200 volts. In a COMFET, that limitation is negated by greatly increasing the conductivity of the epitaxial drain region through the injection of minority carriers due to the additional P- or N-type substrate. Now, in a typical N-channel COMFET, the on-resistance is typically about 1/10 that of a conventional MOSFET with the same chip area, and maybe as little as 1/25 of that value.

DC/DC CONVERTER/INVERTER

The SI7660, from Siliconix, is a highly efficient low-voltage CMOS DC/DC power converter/inverter intended for use as a voltage inverter. However, with the addition of a few components, the inverter circuit can be rearranged to provide several voltage levels at once. Voltage inversion, voltage doubling, supply splitting, and simultaneous inversion and doubling are possible with the device.

The SI7660 operates as a low-cost negative power supply for low-current applications, and it is less expensive, smaller, and uses fewer parts than a dedicated negative power supply. With the addition of only two capacitors and a small-signal diode, the device performs voltage conversion for an input range of +1.5 to +10.0 volts to a complementary output range of −1.5 to −10.0 volts.

The device's output source-resistance is low and allows a drain current of up to 40 mA with a 5-volt input. The low output source-resistance provides high conversion efficiency—95% at 10 mA or 99.7% with no load.

The SI7660 contains a voltage regulator, R-C oscillator, voltage-level translator, four power MOS switches, and a logic network on a single IC. The logic network senses the most negative voltage in the device and ensures that the output N-channel switch substrates are not forward-biased. The oscillator, when unloaded, oscillates at a nominal frequency of 12 kHz for an input supply voltage range of 1.5 to 10.0 volts. The OSC terminal may be connected to an external capacitor to lower the frequency, or it may be driven by an external clock. The LV terminal may be tied to ground to bypass the internal regulator and improve low-voltage operation. At +3.5 to +10.0 volts, the regulator is made operational by leaving the LV pin open.

TRANSISTOR SELECTION GUIDE

Ferranti's 6-page *MOSFET Selection Guide and Cross Reference List* offers design data on complementary N- and P-channel transistors. Arranged as an easy-to-read chart, it provides data on key parameters of over 150 MOSFET-devices in TO-92, TO-39, TO-220, and TO-3 packages.

Included in the guide are low-voltage threshold devices for telecommunications, superfast (1-ns) switching devices, and high-voltage, low-leakage devices for use in test instruments. It also contains key parameters of the various devices covered, among those specifications are: continuous current, I_D; breakdown voltage, BV_{DSS}, and on-resistance, $R_{DS(ON)}$.

The guide also includes a cross-reference to help the user in selecting equivalent or near-equivalent devices.

NEW POWER MOSFET FAMILY

RCA has introduced a new family of power MOSFETs designed to operate directly from logic supply-voltage levels of +5 volts. Called logic-level power MOSFETs or L^2FET devices, they are the first designed specifically to produce full-current outputs with a 5-volt gate drive.

The new L^2FET family includes 32 N-channel devices with current ratings ranging from 1 to 15 amperes, and V_{DSS} ratings ranging from 50 to 200 volts. The devices are available in TO-3, TO-39, and TO-220 packages.

FAST-TURN-OFF SCRs

Recently Motorola introduced a series of SCRs that can turn off ten to twenty times faster than standard types. This latest line of SCR's, available in TO-220 packages, consists of 28 devices included in the MCR2080-A (8 amps) and MCR2150-A (15 amps) series with repetitive reverse voltages ranging from 200 to 800 volts. The maximum turn-off time is 4.0 μs for devices in the MCR2080-A series and 6.0 μs for devices in the MCR2150-A series. The turn-off time for the non-"A" devices is 10 μs.

OP-AMP POWER BUFFER

The LT1010 is a fast op-amp buffer that has unity voltage gain but can amplify an op-amp's output current from typically +10 mA to +150 mA. It makes fast amplifiers less sensitive to capacitive loading, reduces thermal feedback in precision DC coupled amplifiers, and is recommended for a number of fast and slow applications.

When incorporated within the op-amp's feedback loop, the buffer can isolate almost any reactive load. Applications include driving headphones or long cables, powering small motors, operating proportional actuators, and use in operational amplifier based power supplies.

The LT1010 has a 20-MHz bandwidth and a 100 V/μs slew rate. Its quiescent current is 5 mA, while its output can swing ± 10 volts into a 75-ohm load. It maintains its inherent high stability while working into capacitive loads exceeding 1 μF. The device operates from a single supply of at least 4.5 volts or from dual supplies delivering up to a total of ± 22 volts.

Diagrams including such applications as track-and-hold, voltage and current regulators, integrators, and power-supply splitters are included in the 16-page LT1010 data sheet and application note.

Chapter 35

Precision Op-Amps

Precision operational amplifiers are the mainstay of medical and industrial instrumentation. Linear Technology Corporation recently introduced the LT1014 precision op-amp, which is the first improved replacement for such industry-standard, 14-pin quad op-amps as the LM324, LM348, OP-11 and 4156. The LT1014's 50—μV offset voltage, 0.3—μV/°C drift, 0.15—nA offset current, open-loop gain of 8 million, 117—dB common-mode rejection ratio and 120—dB power-supply rejection ratio definitely place it in the precision class.

The offset voltage of the LT1014 is so low that no offset adjustment terminals are provided on the IC. A new and improved output stage draws only 350 μA, can source or sink more than 20 mA, and still retains high voltage gain.

Another member of the family, the LT1013, is the first improved direct replacement for industry-standard, 8-pin dual packages such as the MC1458/1558, LM148 and OP-221. The LT1013s specifications are similar to, but slightly better than, the LT1014s.

Both the LT1013 and the LT1014 can be operated from a single 5-volt supply. The common-mode input range includes ground, and the output can swing to within a few millivolts of ground. Crossover distortion, a characteristic of early single-supply designs, has been eliminated. Linear Technology gives complete specifications for both single-ended 5-volt and ±15-volt operation.

Absolute maximum ratings are: supply voltage: ±22 volts, differential input voltage: ±30 volts, input voltage: equal to the positive supply voltage. Operating temperature range is −55 °C to +125 °C for devices with the AM and M suffixes, and 0 °C to +70 °C for devices with the AC, C, and D suffixes.

USING THE LT1013 AND LT1014

In earlier single-supply designs using such devices as the LM124, LM158, OP-20, and OP-221, if the input goes more than a few millivolts below ground, two kinds of problems may occur, both of which the Linear Technology devices provide protection against.

- If the input falls more than 400 mV below ground, transistors in the input stage saturate and phase reversal could cause lock-up in servo systems. The LT1013/1014 devices include unique phase-reversal protection circuitry that prevents the output from reversing phase, even when the inputs are as low as −1.5 volts.
- If the input falls more than a diode-drop below ground, the device could be destroyed because essentially unlimited current could flow from the substrate ($-V_{SS}$) to the input terminal. The LT1013/1014 devices are protected against that type of failure by a 4000-ohm resistor in series with each op-amp input.

There is one circumstance in which the phase-reversal protection circuitry will not work: when the output of another op-amp in the same package is driven hard into negative saturation. In the LT1013, either op-amp can disable the other's protection circuitry. In the LT1014, amplifiers A and D can affect each other, but amplifiers B and C are completely independent of the former. Similarly, amplifiers B and C can affect each other, but amplifiers A and D are completely independent of them.

AN APPLICATION

Figure 35-1 shows a liquid-flow meter designed around an LT1014 (or two LT1013s). The rate of flow through the pipe is determined from the temperature differential that occurs as liquid flows through a heated section of pipe. The ambient temperature of the liquid is measured by thermistor TI at the input end of the pipe. Thermistor T2 measures the temperature at the output end of the pipe. The temperature difference varies inversely with the rate of flow.

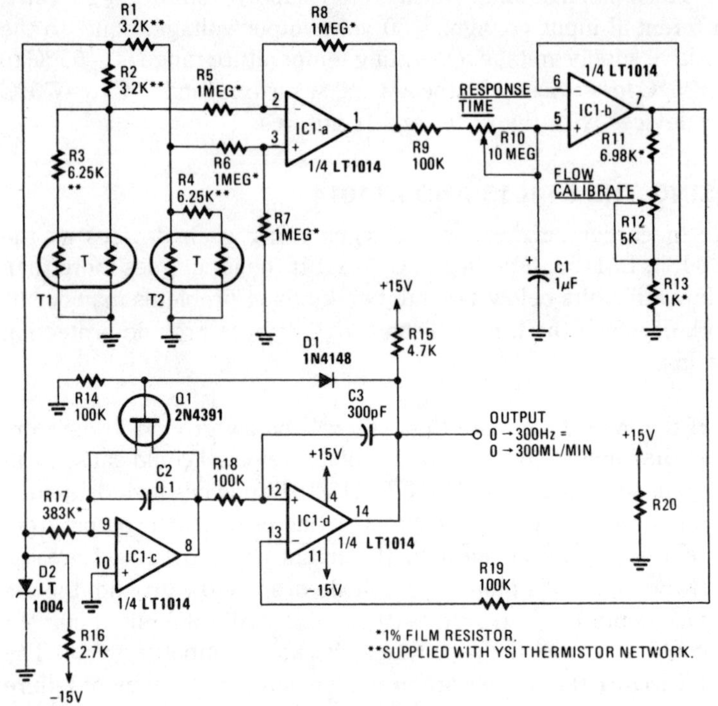

a

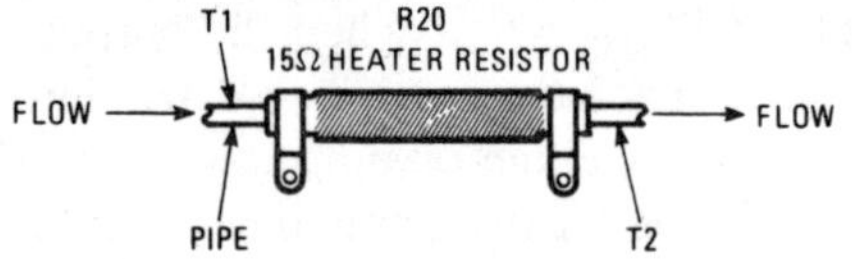

b

Fig. 35-1.

Op-amps IC1-*a* and IC1-*b* amplify the differential voltage developed by the thermistors, and op-amps IC1-*c* and IC1-*d* form a linear voltage-to-frequency converter. The output frequency ranges from zero (for a flow rate of zero) to 300 Hz (for a flow rate of 300 milliliters per minute). The rate of flow can be read directly on any audio-frequency meter.

The 15-ohm heater is a Dale Electronics 25-watt wirewound resistor, type HL-25. Thermistors T1 and T2 are YSI type 44201 thermistor networks available from Yellow Springs Instrument Co., Box 465, Yellow Springs, OH 45387.

Chapter 36

Surface-Mount Technology Comes of Age

Printed-circuit boards were widely used before semiconductors began to appear in consumer-electronics products. Components were almost invariably mounted on the top of the board, with terminal leads protruding through holes to the underside of the board where components were interconnected. That technique remained an international standard as semiconductors replaced vacuum tubes.

With the advent of digital watches, business-card calculators, and portable personal computers, engineers were hard-pressed to cram more circuitry in less space. In addition, the DIP with its increasingly high pin-counts suffers from inductance, resistance, and capacitance problems that adversely affect processing speed.

One solution to those problems is *surface-mount technology,* or SMT, which makes it possible to mount both discrete and integrated semiconductor devices directly on the surface of the PC board. Cost savings are effected due to a reduction in the number of plated-through holes, improved board yields, and higher component density. Components can even be mounted on both sides of the board with little increase in volume. Using SMT, Texas Instruments has achieved space reductions between 46% and 60% over circuits previously assembled on PC boards with discrete components.

SURFACE-MOUNT TRANSISTORS

Several semiconductor manufacturers are now producing transistors in the SOT-23 package—a subminiature three-lead package one-fourth the thickness of the common TO-92 package.

The SOT-23 package was originally introduced in two different heights, but Ferranti Semiconductors has introduced what they call their "optimum profile" package. Ferranti believes that their package eliminates the need for both standard and low-profile packages. The space between the device and the PC board allows for proper cleaning and also allows traces to be run beneath the device, while still permitting the device to be epoxied to the surface of the board. (Since there are no through-the-board leads, surface-mount components must be epoxied to the board before they are soldered. Soldering is usually accomplished by a solder-reflow process.)

Ferranti offers a number of devices for SMT, including both NPN and PNP transistors, ranging from general-purpose, low-noise, and switching devices, to medium-power, high-frequency, and high-voltages types. Also included is a line of switching, Schottky, Zener, and varactor diodes.

Other SOT-23 devices made by Ferranti are the FMMT38 family of NPN Darlington transistors that feature high breakdown voltage (V_{CEO}), and HFE ratings between 1000 and 10,000 at 500 mA and 5 volts (pulsed). Applications range from audio driver and output stages to drivers for lamps, relays, and hammers.

The 6-page "SOT-23 Transistor and Diode Selection Guide" lists key parameters of more than 250 devices for hybrid and SMT applications. Further information on the "optimum profile" package and copies of the selection guide are available from **Ferranti Semiconductors,** 87 Modular Ave., Commack, NY 11725.

A line of fifteen small-signal FETs is now being offered in SOT-23 packages by Siliconix. The devices, available in one P-channel and four N-channel configurations, are characterized as general-purpose amplifiers, low-noise amplifiers, low-leakage amplifiers, high-frequency amplifiers, and high speed analog switches. Prices are approximately 36 cents each in 5000-piece quantities.

SURFACE-MOUNT ICs

As shown in Fig. 36-1 surface-mount ICs are currently being offered in three SMT packages. These are SOIC (*S*mall *O*utline

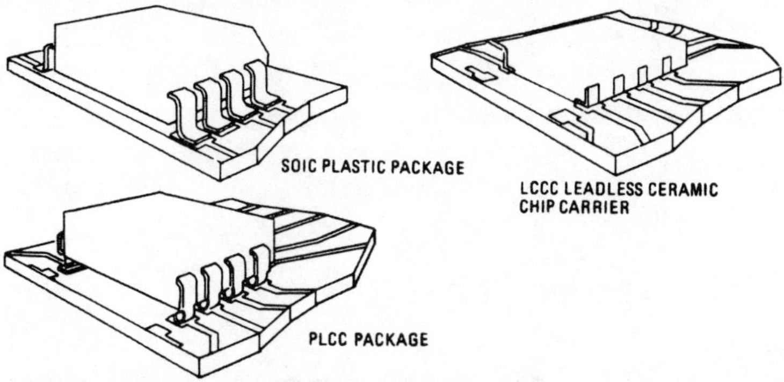

Fig. 36-1.

*I*ntegrated *C*ircuit), the PLCC (*P*lastic *L*eaded *C*hip *C*arrier) and the LCCC (*L*eadless *C*eramic *C*hip *C*arrier).

The SOIC has leads on two sides. Its "gull-wing" leads are bent outward; they lie parallel to the mounting surface. The SOIC comes in 8-, 14-, 16-, 20-, and 24-pin versions. The PLCC is intended to house devices requiring 20 to 68 pins. The PLCC has J-shaped leads spaced equally around the sides of the package. The J-shaped leads are tucked under the plastic body as shown in the middle drawing of Fig. 36-1. The LCCC is similar to the PLCC but is considerably more expensive and is used mainly in military applications.

PLCCs FROM MOTOROLA

The MCM63256 and MCM65256 256K (32K × 8 bit) MOS mask-programmed ROM's are packaged in PLCC's, and the rolled-under "J" leads are compliant, thus reducing temperature-coefficient problems. The 32-lead devices measure 0.450 × 0.550 inches.

The MCM63256 is fabricated using Motorola's silicon-gate HMOS process; those new units are available with access times of 150 and 200 nanoseconds. The MCM65256 is designed for low power dissipation, and uses Motorola's complementary HCMOS technology.

ZENER DIODES IN TO-220 PACKAGES

Motorola now has 10- and 50-watt Zener diodes housed in a TO-220 plastic package that can replace more expensive metal-packaged devices. The 10-watt MZT2970-3015 series offers voltages ranging from 6.8 to 200 volts while the 50-watt MZT3305-3350 and MZT4549-4554 families range from 3.9 to 200 volts.

Chapter 37

Analog ICs and New MOSFETs

Many interesting ICs have recently been developed for audio applications. For example, the MTA1537A is a precision AGC/Voltage-controlled attenuator developed by Aphex Systems. It is specially suited for high-speed precision control of signal level, dynamic range, and phase and amplitude equalization in applications such as high-quality audio controllers, analog computers, precision oscillators, robots, video-effects generators, servo controllers and precision phase detectors. The MTA1537A comes in a standard 14-pin DIP package; typical specifications are:

- 110-dB dynamic range
- 120-dB attenuation
- less than 0.05% THD
- 2-mv control-voltage feedthrough
- 200-MHz bandwidth

Figure 37-1 shows how the MTA1537A can be used as a stereo attenuator. Typical specifications for that circuit include maximum input and output levels of 7.75 volts rms, maximum attenuation of 100 dB over a frequency range of DC to 200 kHz, and a control voltage feedthrough of 5 mV at 100 dB attenuation. Note that a high frequencies, circuit layout will affect attenuation.

The op amps used in that circuit may be LF353s (or equivalents); NE5534's could be used for low-noise applications.

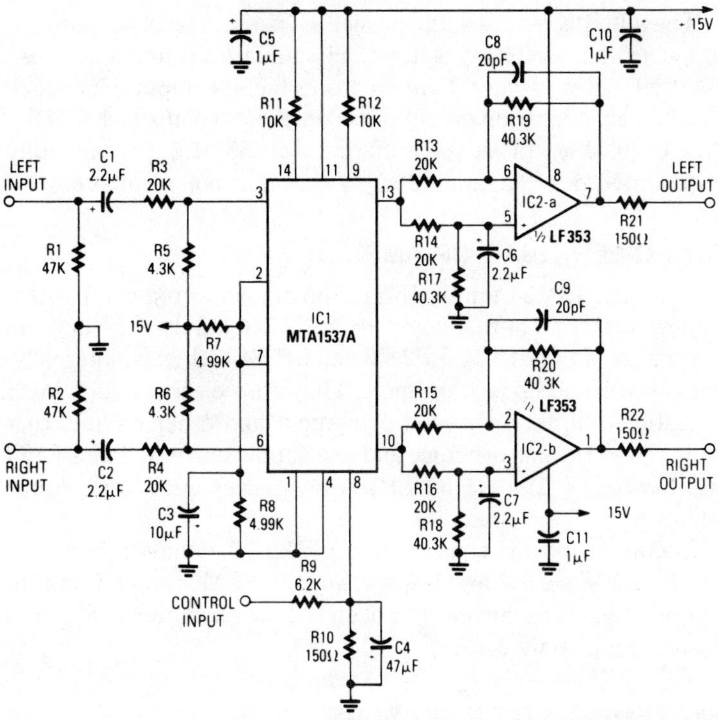

Fig. 37-1.

The MTA1537's data sheets include basic circuit diagrams for applications such as a voltage-variable resistor, voltage-tunable high-pass filter, voltage-variable inductor, voltage-controlled band-reject filter, and a voltage-controlled graphic equalizer.

NEW JFET OP-AMP FAMILY

The MC34080 series is a new generation of JFET-input op amps from Motorola. They are available in single, dual, and quad versions (both compensated and uncompensated), and they offer bandwidth and slew rates as much as four times greater than previously available types.

A combination of JFET and bipolar technologies, along with an all-NPN output stage, has yielded a fully-compensated op-amp family with a gain-bandwidth product of 8.0 MHz and slew rates in excess of 30 V/μs. For greater speed, uncompensated versions are offered with a gain-bandwidth product of 16 MHz and slew rates of 60 V/μs.

The all-NPN output stage provides a peak-to-peak output voltage swing that is 33% greater than standard op amps that use NPN/PNP output stages. Other features include: input impedance of 10^{12} ohms, open-loop output impedance of 30 ohms at 1.0 MHz, THD of 0.01%, phase/gain margins of 55°/7.6 dB for fully compensated devices, and 30 $nV/\sqrt{Hz}$ input noise voltage.

LOW-POWER QUAD OP AMPS

Nationals LP124 family of quad op amps is a pin-compatible, low-power version of that company's LM124 family. The new series consists of the LP124, LP2902 and LP324 high-gain, internally-compensated micropower op amps. They are well suited for CMOS applications, battery-powered equipment and other circuits that require good DC performance and low supply current. The LP124 joins National's LP139 and LP165 low-power quad comparator families.

Maximum power consumption is 125 μA—one-tenth that of National's LM series, and lower than many CMOS amplifiers. In addition, input bias current has been reduced by a factor of ten to a maximum of only 4 nA.

NEW TEMPERATURE SENSORS

The LM34, LM34C, and LM34D make up a series of precision temperature sensors with output voltages linearly proportional to the Fahrenheit temperature scale. Those devices, from National Semiconductor, require no external calibration or trimming to provide typical accuracies of ½ °F at room temperature and $\pm$ ½ °F over a full range of -50 to $+300$ °F. Temperature stability is aided by the LM34's low self-heating, which is typically less than 0.2 °F in still air.

Interfacing to readout or control circuits is easy. The output impedance is only 0.1 ohm at 2 mA and the output is linear at 10 mV/ °F. It can be used with single power supplies or with split (plus and minus) supplies delivering from 5 to 30 volts and draws less than 60 mA. The LM34 can be glued or cemented to a surface and its temperature will be within 0.02 °F of the surface reading.

The LM34 is rated from -50 to $+300$ °F, the LM34C from -40 to $+230$ °F, and the LM34D from $+32$ to $+212$ °F. All are available in TO-46 packages. The LM34C and LM34D are also offered in the TO-92 packages.

Index